DEPARTMENT OF SCIENTIFIC & INDUSTRIAL RESEARCH

NATIONAL PHYSICAL LABORATORY

SYMPOSIUM No. 16

Wind Effects on Buildings and Structures

Proceedings of the conference held at the National Physical Laboratory, Teddington, Middlesex, on 26th, 27th and 28th June, 1963

VOLUME I

LONDON: HER MAJESTY'S STATIONERY OFFICE
1965

© *Crown copyright 1965*

Printed and published by
HER MAJESTY'S STATIONERY OFFICE

To be purchased from
York House, Kingsway, London w.c.2
423 Oxford Street, London w.1
13A Castle Street, Edinburgh 2
109 St. Mary Street, Cardiff
39 King Street, Manchester 2
50 Fairfax Street, Bristol 1
35 Smallbrook, Ringway, Birmingham 5
80 Chichester Street, Belfast 1
or through any bookseller

Printed in England

PREFACE

THIS International Conference, which it is believed is the first to be held on its subject, was organised by the Aerodynamics Division of the National Physical Laboratory with the collaboration of the Building Research Station, the Institution of Civil Engineers, and the Institution of Structural Engineers.

During the three days of the Conference a total of twenty four papers were presented in six plenary sessions, each of which was devoted to one or two aspects of the subject. After a brief opening and introductory session the subjects discussed in each session were broadly: design wind speeds, wind structure, wind loads on full-scale buildings; experimental determination of wind loads in wind tunnels; wind clauses in codes of practice and the response of structures to gusts; aerodynamic oscillation of suspension bridges; and wind-excited oscillations of transmission lines, towers, stacks and masts. The papers presented are reproduced in full, together with an abbreviated account of the discussions.

The participants in the Conference numbered just over 300 and represented 20 countries. The Conference was held in the Glazebrook Hall at the National Physical Laboratory and simultaneous translation of the proceedings into French and English was available to participants through personal radio receivers.

A cocktail party and river cruise were organised as evening social events and on the Saturday a small number of participants journeyed north to inspect the new Forth roadbridge then under construction, and the Forth railway bridge, in a tour organised by the Institution of Civil Engineers.

CONTENTS

SESSION 1

Page

CHAIRMAN: SIR GORDON SUTHERLAND, F.R.S.
Name: National Physical Laboratory, U.K.

Opening address by Mr. Denzil Freeth, M.P. — 3

Paper 24 Introductory review of wind effects on buildings and structures — 9
C. SCRUTON, National Physical Laboratory, U.K.

Discussion on Paper 24 — 24

SESSION 2

CHAIRMAN: SIR GRAHAM SUTTON, F.R.S.
Meteorological Office, U.K.

1 The estimation of design wind speeds — 29
H. C. SHELLARD, Meteorological Office, U.K.

2 The relationship of wind structure to wind loading — 53
A. G. DAVENPORT, University of Western Ontario, Canada

Discussion on Papers 1 and 2 — 103

13 The measurement of wind pressures on tall buildings — 113
C. W. NEWBERRY, Building Research Station, U.K.

Discussion on Paper 13 — 149

4 Détermination de l'action d'un vent turbulent sur les bâtiments et constructions — 151
RAYMOND PRIS, Ingénieur E.C.P., France

Chairman's concluding remarks — 177

Discussion (written) on Paper 4 — 177

NPL SYMPOSIUM 16

WIND EFFECTS ON BUILDINGS AND STRUCTURES

Corrections

VOLUME I

P. 70   Equation (16) to read

$$\frac{n \cdot S(n)}{V_*^2} = \frac{n \cdot S(n)}{\kappa \, \bar{V}^2} = 4 \cdot \frac{x^2}{(1+x^2)^{4/3}}$$

P. 71   Equation (17) to read

$$I_Z = \frac{\sigma}{\bar{V}_Z} = \frac{\left[\int_0^\infty S(n)\, dn\right]^{1/2}}{\bar{V}_Z}$$

$$= \frac{\sqrt{6 V_*^2}}{\bar{V}_Z}$$

$$= 2.45 \sqrt{\kappa} \, \frac{\bar{V}_1}{\bar{V}_Z}$$

P. 198   Last line, equation to read

$$c_p = \frac{p - p_o}{\rho \, V_o^2 / 2}$$

P. 393   Authors initials to read

         R. I. HARRIS

VOLUME II

P.447   Mid-page '$b_{/\!N}$' to read 'b/V'

P.545   Fig. 9(a) delete '(contd)' from title

P.546   Fig. 9(b) insert '(contd)' after 9(b) in title

P.614   '$I\alpha$' to read 'I_α', '$r_\infty \gamma$' to read 'r, γ'

P.633   '$\dfrac{d}{dt}\left(\dfrac{\delta K.E.}{\delta \dot{q}_i}\right)$' to read '$\dfrac{d}{dt}\left(\dfrac{\delta K.E.}{\delta \dot{q}_i}\right)$'

in equation 1.5.12

P.833   '5×105' to read '5×10^5' in paragraph starting with 'MR. WOOD'.

National Physical Laboratory
April 1965

LONDON: HER MAJESTY'S STATIONERY OFFICE

SESSION 3

CHAIRMAN: MR. E. N. UNDERWOOD
Institution of Structural Engineers, U.K.

Paper		Page
15	Model law and experimental techniques for determination of wind loads on buildings NIELS FRANCK, Denmark	181
6	Effects of velocity distribution on wind loads and flow patterns on buildings W. DOUGLAS BAINES, University of Toronto, Canada	197
	Discussions on Papers 15 and 6	224
5	Préparation des essais sur maquettes de bâtiments au laboratoire aérodynamique et applications à la vraie grandeur RAYMOND PRIS, Ingénieur E.C.P., France	227
22	Exécution en tunnel aérodynamique d'essais sur maquettes de bâtiments en rapport avec les mesures faites sur constructions réelles P. E. COLIN and R. D'HAVÉ, *respectively,* Training Centre for Experimental Aerodynamics, Belgium, Bureau de Controle pour en Sécurité de la Construction en Belgique	255
21	Model simulation of wind effects on structures R. E. WHITBREAD, National Physical Laboratory, U.K.	283
	Discussions on Papers 5, 22 and 21	303

SESSION 4

CHAIRMAN: DR. F. M. LEA
Building Research Station, U.K.

19	Les Règles Françaises 1963 définissant les effets du vent sur les constructions N. ESQUILLAN, Institut Technique de Bâtiment et des Travaux Publics, France	309

Paper		Page
14	Proposed code of practice for wind loads for Denmark MARTIN JENSEN and NIELS FRANCK, Denmark	333
	Discussions on Papers 19 and 14	352
9	The buffeting of structures by gusts A. G. DAVENPORT, University of Western Ontario, Canada	357
18	The response of structures to gusts R. I. HARRIS, Electrical Research Association, U.K.	393
	Discussions on Papers 9 and 18	422
	Written discussions on Papers 19, 14, 9 and 18	424

SESSION 5

CHAIRMAN: SIR ALFRED PUGSLEY, F.R.S.
University of Bristol, U.K.

10	Instabilité des ponts suspendres dans le vent - expériences sur modèle réduit Y. ROCARD, University of Paris, France	433
11	Aerodynamic effects on suspension bridges A. SELBERG, Norges Tekniske Høgskole, Norway	461
20	A summary of laboratory and field studies in the United States on wind effects on suspension bridges G. S. VINCENT, Bureau of Public Roads, USA	487
17	The use of models to predict the oscillatory behaviour of suspension bridges in wind D. E. J. WALSHE, National Physical Laboratory U.K.	517
	Discussions on Papers 10, 11, 20 and 17	555

SESSION 6

CHAIRMAN: PROFESSOR J. S. PIPPARD, F.R.S.
Institution of Civil Engineers, U.K.

Paper		Page
12	The status of the conductor galloping problem in Canada A. D. HOGG and A. T. EDWARDS, The Hydro-electric Power Commission of Ontario, Canada	561
23	Aeroelastic galloping in one degree of freedom G. V. PARKINSON, University of British Columbia, Canada	581
7	Research study on galloping of electric power transmission lines A. S. RICHARDSON, J. R. MARTUCCELLI and W. S. PRICE, Research Consulting Associates, USA	611
8	Aerodynamic properties of the Severn crossing conductor D. J. W. RICHARDS, Central Electricity Research Laboratories, U.K.	687
	Discussions on Papers 12, 23, 7 and 8	766
3	An experimental study of aerodynamic devices for reducing wind-induced oscillatory tendencies of stacks K. NAKAGAWA, T. FUJINO, Y. ARITA, and T. SHIMA, University of Osaka Prefecture, and Mitsubishi Shipbuilding and Engineering Co. Ltd., Japan	773
16	On the wind-excited oscillations of stacks, towers, and masts C. SCRUTON, National Physical Laboratory, U.K.	797
	Discussions on Papers 3 and 16	833

APPENDIX

List of participants 837

SESSION 1

Chairman:

Sir Gordon Sutherland, F.R.S.
(Director, National Physical Laboratory, U.K.)

OPENING ADDRESS

by

Mr. Denzil Freeth, M.P.
(Parliamentary Secretary to the
Minister for Science, U.K.)

OPENING ADDRESS

by

Denzil Freeth, M.P.
(Parliamentary Secretary to the Minister of Science)

I do not know how many of you in the days of your youth have ever read any of those so-called "Gothic" novels produced by the romantic novelists of almost every country in the early years of the 19th century. In each of these there would almost certainly be at some time a scene depicting a ruined castle or an isolated ancient mansion in which as night fell deeds of darkness were going to take place: rich relatives were to be poisoned, virgins were to be abducted, those who knew more than they should were to be silenced by torture or by death.

If you had read more than one of those novels you would be prepared to bet your bottom dollar that at the time when these dreadful deeds were to take place the sky would be darkened by a covering of scurrying clouds, and across the plain, or round the crags and the turrets and the chimney pots, a wind would whistle which would keep the traveller at home and provide a fitting musical accompaniment to the horrors that were to take place indoors.

But most of us, whether we read such novels in our youth or not, have, I hope, no direct knowledge of such deeds. We have, however, all had the experience on a winter's night of sitting, if not by our firesides, at any rate by the radiators of our central heating system, secure in the cosy warmth of home or office, listening to the wind rushing across the land. I think, however, that we would not have sat with such feelings of comfort if we had not been pretty certain that the building in which we were sitting was likely to resist the onslaught of the gale.

It is unfortunately true that buildings do not always, and in the past certainly have not always, resisted the onslaught of wind. In this country, for example, in the Sheffield area in 1962, especially severe gales caused damage to buildings which has been estimated at no less than £20 million in value; and there have been occasions in previous years when roofs on housing estates have been lifted by the blast and thrown to the ground.

In addition, there has been across the years the hazard to bridges, which have been overturned, and to tall structures, such as pylons, which have been blown over, a hazard which has occurred not merely in time of gale but also because of the steady load due to the wind. All these failures of structure in the face of nature have resulted in a waste of economic resources and of man-hours when those buildings have had to be restored, and in a loss of comfort and amenity, if not of life, for human beings to which no monetary figure can be put.

All this makes it obvious even to the layman that the more that can be known about the effect that wind has on buildings and structures, the more likely it is that we shall be able to build such structures as will withstand the onslaught of nature, and also do so - because this is important - without waste and without having overestimated the strength needed or the design speed likely to be encountered, with the consequent waste of resources and the consequent increase in costs.

It is, therefore, with a feeling of very great pleasure that I come today to open your Conference. I am told that it is the first Conference ever held in the world which is devoted exclusively to the consideration of wind effects upon buildings. I believe that those who have organised this Conference deserve the greatest praise for their efforts. These bodies are the Institution of Civil Engineers, the Institution of Structural Engineers, the Building Research Station of the Department of Scientific and Industrial Research, for which the Minister for Science is responsible to this nation, and the Department's National Physical Laboratory, in whose premises we are now meeting. I am very grateful to the staff of this Laboratory - and I am sure everybody is - for undertaking the bulk of the work of organising this Conference.

I rejoice, too, that this is a Conference which is bringing together people of differing fields of knowledge and differing backgrounds - from the research worker, both in the universities and in research establishments, to the practising engineer. I am glad, too, that this Conference is an international one. Some 24 papers are being presented, not merely by people from the United Kingdom but also by very distinguished people from eight other countries; and visitors from some 10 or so other countries have come here to participate in the discussions and will, I know, make a great contribution to the success of the next few days.

I think also that this Conference is particularly timely because of the increasing importance which is being accorded today, and rightly accorded, to the effect of wind in the design of modern structures. In many countries of the world we are today building higher than ever before, and because of new materials and new knowledge, architects are able to create new, novel and exciting architectural designs. New methods of fabrication and new materials for use in buildings all mean that we must learn more about the effects of wind if the creative power of the architect and the work of the manufacturer of new materials and the designer of new methods of fabrications are, indeed, to have a worthy and a lasting memorial.

Although the problem of wind has been with us since, I suppose, the beginning of time, it is only within the last 70 years or so that wind loading has been the subject of studies in wind tunnels, and even then in the past these studies have often related to specific structures rather than to the more general and systematic investigations which alone can provide a code of practice for building which the building engineer can use as the source of his information. The present trend towards higher and more slender buildings and the use of modern engineering materials and methods of construction have, naturally, led to a demand for more refined structural analysis, and this demand has coincided with two other developments which I think you will find of interest in your discussions over the next few days. First, wind tunnels and other facilities for studying wind effects have become more freely available in recent years; and, secondly, the systematic study of meteorological observation and meteorological research enables more knowledge to be obtained of the likely design load which a building must sustain. The estimation of the design speed of wind is, therefore, of the first importance, and, appropriately enough, it is the first topic which you will be discussing under the chairmanship of the Director of the United Kingdom's Meteorological Office this afternoon.

The renewed interest in this subject has resulted in a number of countries, including the United Kingdom, deciding to revise their codes of practice for building in respect of defence against wind. I note that among the papers submitted to the Conference there are two describing the proposed wind loading clauses for the Danish and French codes of practice. This is an international problem, and in putting the results of research into building practice we can all learn from one another: we shall certainly study with great care the experience and the proposals of our Danish and French friends.

Research into this problem can, naturally, take place in a variety of environments. There will always be, I hope, research in universities. There will also be research in laboratories run by Government, such as the National Physical Laboratory where we now are and where you can see an exhibition and demonstrations of the models and techniques which are used here for investigating wind effects on structures. Research also takes place in this country at the Building Research Station, which has developed special pressure measuring instruments for use on buildings, and a number of these are being installed, and, indeed, have been installed, on high buildings recently erected in London, including State House, the headquarters of the Department of Scientific and Industrial Research.

Then, in this country there are co-operative research associations supported in the main by industry itself, but with grants from the Government. I see that on Thursday a member of the Electrical Research Association is to describe work which forms part of a continuing and extensive, but not yet completed, study of the special problems of wind effects on very tall transmission masts and towers.

But there must also be research which is conducted by industry itself, or at any rate which is commissioned and financed by industry, whether that research is actually to be done in universities, or in Government laboratories, other research laboratories, or carried out in the laboratories of industry itself. If we are to have a broad and free-flowing channel of communication between the research worker in the laboratory on the one hand and the architect and the engineer who actually construct the building on the other, then it is essential that at every stage there should be close contact and an ability not only intellectually to understand but practically to achieve the results of the research and development teams.

In the name of Her Majesty's Government, I would like to welcome you all here. I would like to wish you every success in your discussions. I hope that these, together with the papers which are to be presented, will be published, and I look forward to reading the publication.

I thank you for having done me the honour of asking me here to open this Conference, and that, with great pleasure, I now do.

PAPER 24

INTRODUCTORY REVIEW OF WIND EFFECTS ON BUILDINGS AND STRUCTURES

by

C. SCRUTON
(National Physical Laboratory, U.K.)

INTRODUCTORY REVIEW OF WIND EFFECTS ON BUILDINGS AND STRUCTURES

by

C. SCRUTON
(National Physical Laboratory)

1. INTRODUCTION

THE purpose of this Paper is to review the present position with regard to wind effects of buildings and structures, and to mention briefly the various topics for discussion at the Conference. Most of the information on which this review is based is necessarily related to experience in the United Kingdom, but the problems are common to all parts of the world and, except perhaps for the hurricane zones, it is not expected that the situation will be significantly different in other countries. Attention is restricted to the wind effects influencing the static and dynamic stability of structures and of structural members. Other effects of wind on buildings, such as the generation of noise, the promotion of fire, and the penetration of rain, are not considered.

With the above exceptions the action of wind can be classified into the *static effects* and the *dynamic effects*. The former refer to the steady (time-average) forces and pressures tending to give the structure or its component members a steady displacement, while the latter refers to the tendency to set the structure oscillating. Figs. 1 and 2 are graphic evidence of the static effects. Fig. 1 shows gale damage to lightweight monoslope roofing. Fig. 2 shows an electricity pylon which was destroyed by gale. Both of these incidents occurred in recent years.

The complete destruction of buildings or structures by wind action is of rare occurrence; but local failures especially to the roofs and cladding of buildings, are much more common and in the aggregate much more costly.

Galileo and Newton in the 17th century are reported to have considered wind loading on buildings, but it was not until late in the 19th century that efforts were made to determine a wind loading to be used for design purposes. The Tay Bridge disaster in 1879 is reputed to have prompted bridge engineers to make allowance for wind loading, and a horizontal wind load of 56 lb/ft^2 (274 kg/m^2) of projected area was adopted for the design of the Forth Railway Bridge as a result of some outdoor tests on flat plates. In subsequent years this value has been reduced considerably as more reliable aerodynamic data and information on natural winds accumulated. The improvement in our knowledge of wind loading is associated to a large extent with the use of the wind tunnel for measurements of aerodynamic forces and pressures. Mention might be made of the pioneer wind-tunnel researches of Irminger in Denmark at the end of the 19th century and of Eiffel in France. Since these early uses of the wind tunnel for the determination of wind pressures on buildings there has been a steady increase in its use both for the systematic investigation of various building shapes in common use, and for the determination of the pressure distributions and overall wind forces on specific building projects. The results of the former type of investigation are suitable for incorporation in the wind loading clauses of the codes of practice for buildings, and most of such investigations were undertaken for this purpose. The detail in which wind loading data are presented in the various national building codes varies widely from country to country, but even the more comprehensive compilations of data, such as that of the Swiss Code, are rarely adequate for the design of major projects, especially if these incorporate novel features. In the absence of sufficient existing information, either from the codes or from other similar compilations of data, recourse can be made to a wind-tunnel investigation. Care in the design of the wind-tunnel experiments and in the processing of the results is necessary for the results to be applied with confidence to the full-scale structure. The experimental methods used in wind-tunnels and the application of the result to full-scale, form the main theme of the papers to be presented in Session 3.

From time to time the wind clauses of the building codes need to be revised to bring the information up-to-date and more adequate for the requirements of the advances in structural design and of modern methods of construction and fabrication. The British Code is at present under revision and Paper 19 by M. Esquillan, and Paper 14 by Dr. Jensen and Mr. Franck give details of the proposals for the new French and Danish codes respectively.

The uncertainty in the estimation of wind loading results largely from the difficulties in the prediction of the maximum wind conditions to which the structure will be exposed. Considerations of safety demand that these must not be underestimated while those of economy do not permit a gross overestimation. Some of the factors influencing the choice of a design wind speed will be mentioned later in this Paper. As a result of the increase in the amount of wind data available, and recent analyses of these, it is believed that more reliable predictions of design wind speeds can now be made, although considerable uncertainties remain. Valuable contributions to our knowledge and understanding of the structure of the wind (especially of that over the British Isles), as it affects the choice of design wind speeds, have been made by Mr. Shellard and Professor Davenport, and these are contributed to this Conference in Papers 1 and 2. Professor Davenport's suggestion of referring design wind speeds to the gradient height, and of relating the speed at lesser heights by factors dependent on the terrain, is especially interesting and, in principle at least, would appear to overcome some of the difficulties associated with the use of the surface wind approach. The dependence of the shape of the velocity profile on the roughness of the terrain over which the wind passes is the essential concept of Dr. Jensen's model law which is expounded in Paper 15 by Mr. Franck.

The design wind loads thus depend upon the choice of the design wind speed, and a knowledge of the non-dimensional force or pressure coefficient which are usually derived from wind-tunnel tests. In the absence of aerodynamic scale effects in extrapolating model results to full-scale, the relationship is simply

$$F = \tfrac{1}{2}\rho V^2 C$$

where
- F is the force per unit area
- ρ is the air density
- V is the design wind speed

and
- C is the force or pressure coefficient.

The adequacy of this approach for the determination of wind loads can only be judged by full-scale experience and in particular by comparison of model results with measurements on actual buildings. The requirements for obtaining the full-scale data are discussed by Dr. Pris in Paper 4. The information is difficult to obtain and very little is as yet available. However, such measurements are in hand in a number of countries and in the United Kingdom the Building Research Station has undertaken to mount specially developed pressure gauges on the walls of a number of the tall buildings in London. Some preliminary results of this investigation are described in Paper 13 by Mr. Newberry.

The dynamic effects of wind have been studied for a much shorter time than the static effects. This is because dynamic effects were not so evident in the older structures but also to some extent because the contribution of the dynamic action of wind to structural failures of the past was not always recognized. The present day interest in dynamic effects started in 1940 with the collapse of the suspension bridge over the Tacoma Narrows due to oscillations set up by wind. The oscillations of overhead transmission lines were noted and discussed before this date but the problems attracted very little research effort. The past history of suspension bridges reveals that many were damaged or destroyed by the oscillatory effects of wind, and these disasters may account for the unpopularity of long-span suspension bridges in Europe which has persisted until recent times. An early example is that of Telford's original suspension bridge over the Menai Straits. Wind excited oscillations damaged this bridge on several occasions in the early part of the century. In 1836 one span of a chain pier at Brighton was broken by torsional oscillations of the same antisymmetric mode which destroyed the Tacoma bridge. (Compare Figs. 3 and 4.) Despite these and other incidents the wind-excited oscillations of suspension bridges were not studied until after the Tacoma disaster. At the present time it would seem that all major suspension bridges are designed with the aid of the guidance provided by a wind-tunnel investigation of its aerodynamic stability properties. Discussions of this topic occupy the whole of Session 5 of the Conference with papers by investigators in France, Norway, United States and United Kingdom.

Wind-excited oscillations are very sensitive to the amount of structural damping, for if oscillation amplitudes are to be maintained or to grow the energy input from the windstream must equal or exceed the energy dissipated by the structural damping. More efficient design and modern methods of fabrication, such as the increased use of welding and fitted bolts for making joints, tend to reduce the structural damping and consequently to increase the liability to oscillate in wind. For this reason the incidence of wind-excited oscillations of stacks and towers has increased in recent years, and there has been a corresponding increase in the attention given to the problem by the research establishments. Paper 3 by Professor Nakagawa and Paper 16 by the writer discuss these problems and the wind-tunnel techniques used to investigate them. Reference is made in both papers to types of aerodynamic spoilers which have been developed to suppress the vortex-type excitation to which the wind-excited oscillations of many structures can be attributed.

The slow large amplitude vertical oscillations of transmission lines, often referred to as "galloping" have been the subject of somewhat sporadic observation and investigation for the past 40 years. A simple aerodynamic explanation of the galloping of transmission lines with their cross-section distorted by ice-formation was given by Den Hartog in 1930. Since then more complete analyses have been presented. In Paper 23 Professor Parkinson briefly reviews his work on the galloping of rectangular sections in a single degree of freedom with special emphasis on the fluid motion aspects of the mechanism producing the excitation. An experimental investigation and a theoretical analysis of systems with several degrees of freedom is described in Paper 7 by Mr. Richardson. The galloping phenomenon is naturally of concern to those countries where the climatic conditions are such that icing of the conductors is of frequent occurrence. It has been observed and investigated extensively in Canada; the present position in that country is given in Paper 12 by Dr. Hogg and Mr. Edwards. Recent experience, however, has demonstrated that icing is not an essential condition for galloping instability. The 5310 ft span power crossing over the River Severn galloped without any ice forming on the conductors. The explanation for these oscillations, as suggested in Paper 8 by Mr. Richards, is to be found in the peculiar aerodynamic effects introduced by the lay of the conductor strands when the conductor is yawed to the wind.

So far this review has been confined to discussion of the effects of winds steady in speed and direction. Natural winds, however, are turbulent and are continually fluctuating. The assumption usually made hitherto is that these fluctuations are so irregular and random that the response of a structure will not differ from that due to a steady wind of the same average speed. The validity of this assumption has been questioned in recent years. The work of Professor Davenport on the response of flexible structures to a spectrum of horizontal gustiness, described in Paper 9, suggests that the displacements produced can be several times as large as those due to a steady wind of the same average speed. Although this result may apply only to structures with a long natural period of oscillation, its implications are far-reaching. For very long or high structures, which are probably the most sensitive to the influence of the gust spectrum, the analysis is further complicated by the finite dimensions of gusts and the necessity to examine the correlation of the gust speeds along the length of the structure. A research team at the Electrical Research Association is currently studying these effects with special reference to the problems of very tall masts and towers. Some aspects of this work are described by Mr. Harris in Paper 18. Further development and much more data will be required before the influence of the gust spectrum can be fully analysed and simplified for general application. In the meantime most structures will continue to be designed on the basis of having to withstand the loads due to steady winds, and for this the assessment of realistic maximum wind speeds for use as the design wind speed is of first importance.

2. THE DESIGN WIND SPEED

The primary sources of the information on which design wind speeds are based are the wind records at many meteorological stations made over a number of years or, in special cases, the recordings made at the site of the proposed construction. These data are treated statistically to give a probable maximum speed likely to occur within a specified period of years. Some of the considerations which enter into the choice of a design wind speed are:

(i) The anticipated lifetime of the structure

The probability theory used in the analysis of wind data yields increased maximum wind speeds as the life period for the structure is increased.

(ii) Duration and volume of gusts

It will be evident from an examination of a continuous recording of the instantaneous speed of natural winds that maximum speeds will vary according to the gust period considered, and will be higher for the shorter than for the longer gust period. It is therefore important to consider what period should be used for the design wind speed. Gusts of a few seconds may suffice to build up local pressures on cladding and individual structural members, but a longer period may be required for the gust to envelope, and for the pressure to build up, over the whole building or structure. Little is known of the forces on a body due to accelerating winds. Some wind-tunnel measurements on circular cylinders and flat plates show that the wind load does not immediately attain its final value for the steady wind, but indeed that it may exceed this value for a brief initial period during the time the wake takes to become fully established. Also the duration of the wind force due to a single isolated gust regarded as an impulsive force is significant in relation to the dynamic response of the structure. Reference has already been made to the response of a structure to the gust spectrum. It would be a convenient simplification if it should prove that the effect of the gust spectrum can be expressed by a simple factor to be applied to the steady design wind speed. Such a factor, it would seem, would be dependent on the natural frequency of the structure.

(iii) Variation of wind speed with height

For the higher structures an accurate knowledge of the increase of wind speed with height is of paramount importance. It is now well-known that the velocity profile for maximum speeds will vary with the gust period and with the terrain.

(iv) Vertical inclination of the wind

For some applications, as for example for bridge design, the vertical inclination of the wind is of interest because of its influence on both static and dynamic stability.

3. DETERMINATION OF THE SHAPE AND PRESSURE COEFFICIENTS

The shape and pressure coefficient data in current use have been derived almost entirely from wind-tunnel experiments. While measurements in wind-tunnels may seem to be a fairly straightforward procedure, certain aspects present difficulties and may lead to errors in the application of the results to the full-scale unless the correct techniques are adopted. Session 3 of this Conference is devoted to discussion of wind-tunnel techniques. It will be noted that none of the Papers describe techniques for simulating gusting winds. Some of the problems associated with wind-tunnel testing are mentioned below.

(i) Representation of the ground effect and of the velocity gradient

In a closed-jet tunnel the floor of the wind-tunnel may be used to represent the ground. In an open-jet tunnel it is necessary to introduce a ground plate. The model and ground plate arrangements are discussed in Paper 5 by Dr. Pris and in Paper 22 by MM Colin and d'Have. The influence of the velocity profile on the pressure distribution over buildings is the subject of Papers 15 and 6 by Mr. Franck and Professor Baines. There are a number of ways by which a velocity profile can be introduced into the tunnel wind stream; the Paper 21 by Mr. Whitbread mentions these; but the reproduction of a specified profile is usually difficult and tedious. A possible exception to this is the method suggested by Dr. Jensen and described in Mr. Franck's paper.

(ii) Aerodynamic scale effect

Model tests can only be carried out at smaller values of the Reynolds Number than those obtaining on full-scale. The aerodynamic scale effect is referred to in several of the papers, in particular those by Dr. Pris and Mr. Whitbread. Fortunately, except for structures of rounded cross-section, the effect is not very large.

(iii) Influence of architectural features

On small models architectural features cannot always be represented in sufficient detail to ensure the accurate determination of local pressure distributions. Small details such as the amount of overhang of a roof or the presence of a parapet or chimney stack, can significantly influence the wind pressures. The pressure distributions shown in Fig. 5 illustrate the very marked effect due to the addition of a parapet to a hyperbolic paraboloid roof. Suction pressures are reduced because the parapet raises the core of the conical vortices formed at the edges of the roof away from the roof surface.

(iv) Determination of internal pressures

The net wind force on the cladding of a building results from the combination of pressures acting on the internal and the external surfaces, and the contribution of the internal pressure is often substantial. Even for a building with nominally impermeable cladding, wind can produce significant change in the internal pressure. Because the model construction cannot represent the full-scale construction, the model cannot be expected to yield values of the internal pressure applicable to the full-scale. However, this difficulty does not arise for buildings with large openings in the cladding which can be adequately represented on the model.

(v) Tunnel wall interference effects

Both the overall wind forces and the pressure distribution over a model in a closed-jet wind-tunnel are affected by the constraint on the airflow due to the proximity of the tunnel walls. In his discussion of this effect Mr. Whitbread recommends that the size of the model in relation to that of the tunnel working section should be severely limited. These remarks do not apply to the open-jet type of wind tunnel for which the corrections for blockage are usually assumed to be negligibly small.

4. THE OSCILLATORY EFFECTS OF WIND

The problems of the wind-excited oscillations of structures cannot be treated theoretically without the use of at least some experimental data, and they provide some of the more interesting wind-tunnel investigations in which the design and construction of the flexible model to simulate

the full-scale is of first importance. For steady winds the principles of
model construction to obtain dynamic similarity with full-scale are now
well-established but the large size of the structures concerned, as for
example long-span suspension bridges and tall transmission masts, often
makes for difficulties in reproducing the required detail of aerodynamic
shape and of mass and stiffness on small models. A serious difficulty in
use of model experiments for the purpose of predicting the behaviour of
the full-scale structure is that the oscillatory phenomena are acutely
dependent on the amount of structural damping and, while the structural
damping on models can be determined and varied over a wide range, there
appears to be no reliable method for determining the damping of the actual
structure whilst in the design stage, and measurements on the completed
structure are also often difficult to make. A great many structures are
inherently unstable aerodynamically and depend on their structural damping
for their freedom from wind-excited oscillations. It is not uncommon with
two identical full-scale constructions, of tall stacks for example, for
their oscillatory behaviour in wind to differ to the extent that one
oscillates while the other remains satisfactorily stable. This difference
in behaviour can only be attributed to differences in the amount of structural damping which might well arise from different foundation conditions.

While there have been a small number of systematic investigations of
the aerodynamic stability of structural shapes in general use, and still
more ad hoc investigations relating to specific projects, the aerodynamic
data so far obtained remains insufficient, even for the simplest structural
cross-sections such as the circle and the rectangle, to enable the aerodynamic excitations and hence the aerodynamic stability of a structure, to
be calculated with any degree of certainty.

The response of a structure to the spectrum of gusts present in the
natural wind has already been mentioned. Similar considerations also apply
to the response of a structure immersed in the turbulent wake of another.
It has been noted, for example, that the leeward of a pair of stacks often
oscillates with greater amplitudes than those of the windward stack.
Oscillations produced by a turbulent wake are termed buffeting oscillations. Some wind-tunnel observations have shown that buffeting might
become a serious consideration in certain situations as, for instance,
when a long-span bridge of low natural frequency is to be placed in close
proximity and approximately parallel to another. Suspension bridge oscillations due to wind are of rare occurrence and, with the attention now
given to the aerodynamic stability of suspension bridges in the design
stage, it can be expected that such incidents will become even less
frequent. On the other hand reports of stack and tower oscillations are
becoming more frequent, as are vibrations of structural members. The

towers for the suspension bridge now under construction over the Firth of Forth oscillated in a bending mode with double-amplitude of about 7 feet whilst they were in the freestanding condition and before their interconnection by catwalks etc. There had been no reports of suspension bridge tower oscillation previous to this incident, and its appearance in this instance can be attributed to a lighter weight construction than was usual hitherto. It is of interest to note that wind-tunnel tests carried out subsequently on an aeroelastic model of the towers proposed for the Severn Bridge, reproduced the behaviour observed on the Forth Bridge towers.

No paper to be presented at this Conference refers specifically to wind effects on large radio telescopes. Because of their size, the wind loads are very large, especially when a solid sheeted-in reflector bowl is used. An accurate assessment of the forces and moments acting on the bowl due to wind is therefore an essential requirement for design purposes. There are also several potential ways in which aerodynamic oscillations can be excited. A potential instability in pitching motion of the bowl of the Joderell Bank radio telescope was found in model tests, and resulted in the designers incorporating a special damping device. With some methods of support coupled oscillations, analogous to the flexure-torsion flutter of aircraft wings, would also seem a possible type of instability. The wind effects on radio telescopes are to be the subject of a 4-day conference to be held in New York in September of this year.

5. CONCLUSION

The present day trends in the design and construction of buildings and structures increase the importance of both the static and the dynamic effects of wind.

For design purposes the chief difficulty arises in the estimation of the most severe wind conditions which the structure will be required to withstand, and there is need for more data on natural winds, especially in towns and for heights up to at least 1000 feet.

Wind-tunnel techniques have been developed in a number of countries for the determination of wind loads and for the investigation of wind-excited oscillations.

While wind-tunnel investigation will be required to provide the aerodynamic data for special projects, the aim of the building codes of practice should be to incorporate wind clauses giving the most complete and reliable information on wind effects possible for general application.

It is hoped that this Conference will help in establishing liaison between the interested organizations in the various countries so that research can be co-ordinated and the results disseminated on a wider and more systematic basis than has been possible hitherto.

* * * *

At the conclusion of the presentation of this Paper a short ciné-film depicting the ovalling oscillations of a stack at the Leicester Power Station was shown. The film provided a graphic demonstration of the effects of vortex-excitation. The remedy for these oscillations is the very simple one of fitting stiffening rings to the top of the stack.

Fig.1. Light-weight roofs lifted by wind.

Fig.2. Electricity pylon blown down by gales.

Fig.3. Torsional oscillations of the Brighton Chain Pier 1836.

Fig.4. Torsional oscillations of the Tacoma Narrows Bridge.

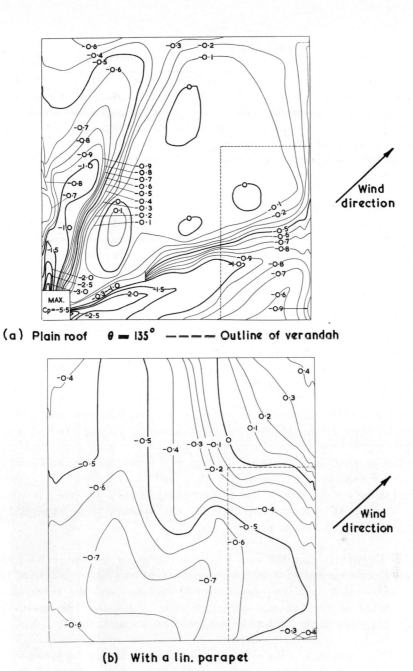

Fig.5. Cp distribution on a hyperbolic paraboloid roof.

DISCUSSION ON PAPER 24

MR. SCRUTON answered a number of questions on the ovalling oscillations of the stack shown in the ciné-film. He thought that the reason why one stack oscillated while another apparently similar stack did not was perhaps that one had corroded more than the other rather than that the structural damping of the two stacks differed because of differences in the foundation conditions. The wind was across the line of the stacks so that wake buffeting did not account for the phenomenon.

DR. PRIS said that wind-tunnel tests to provide data for the wind clauses of the codes of practice were made with very simplified models in which minor constructional details could not be represented. In some cases, however, as Mr. Scruton has shown, the minor details can be very important, and so there remains a great deal of work to be done, especially on the effect of these minor details of construction.

MR. O'HEA said that while he appreciated that the primary object of the Conference was to discuss the structural aspects of wind effects, he would like to stress that wind was of importance in relation to the comfort of those inside the building through its effect on the proper operation of the heating and ventilation services.

DR. PRIS was much interested in ventilation problems and made the distinction between forced and natural ventilation. He had studied natural ventilation problems on models in wind-tunnels for some time, and in particular he mentioned those of a large new steel factory at Dunkirk where it was essential to remove the dust and toxic gases efficiently. Model tests in a wind-tunnel had proved satisfactory for such investigations. It was, of course, necessary for the studies to be made while the building was in the design stage.

DR. KAPLAN pointed out the world-wide interest in the subject of the Conference. He agreed with Mr. Scruton that it was important that the liasons afforded by the Conference should continue and the research work on the subject in the various countries be co-ordinated. He asked for specific suggestions as to how this should be accomplished.

MR. LISTER (written). There are no reports of damage incurred by tubular flood lighting towers manufactured by my firm. Some of these were erected in the Sheffield, Bradford and Nottingham areas where severe damage, or complete collapse of other types of towers were experienced. The tubular towers were designed in accordance with B.S. 449 and the wind load requirements according to Chapter 5 of the Code of Practice on Buildings, the design wind speed taken as 80 miles an hour.

Experience would indicate therefore that a few failures should not be taken as conclusive evidence for the inadequacy of the present design standards especially where tubular construction is concerned. It would also be important to investigate the true nature of the failures experienced particularly to confirm that these were not related to departures from existing design standard requirements.

SESSION 2

Chairman:

Sir Graham Sutton, F.R.S.
(Director, Meteorological Office, U.K.)

PAPER 1

THE ESTIMATION OF DESIGN WIND SPEEDS

by

H. C. SHELLARD
(Meteorological Office)

THE ESTIMATION OF DESIGN WIND SPEEDS

by

H. C. SHELLARD, B.Sc.

(Meteorological Office)

1. INTRODUCTION

WHEN designing any structure to be erected out of doors one of the chief factors that the engineer has to take into account, affecting both cost and safety, is the maximum load that is likely to be imposed on it by the wind. It is not therefore surprising that as increasing numbers of large buildings and other structures have been planned and erected during the post-war years, so meteorological services have received more and more requests for data and advice concerning maximum wind speeds, particularly as many of the structures involved have been novel in design, larger than anything previously contemplated or sited in very exposed places.

Formerly it was often considered sufficient to use as a basis for design the highest wind speed that had actually been recorded at the meteorological station nearest to the place concerned, but this was usually an uncertain guide for many reasons. This "highest recorded speed" depends on such factors as the length of the record available, the height of the anemometer above the ground and the nature of its surroundings. If the maximum gust is being considered then it is also dependent on the sensitivity of the anemometer; if it is the highest mean wind speed then the averaging period needs to be specified. Nowadays many civil engineers are familiar with these complications and to an increasing extent they seek meteorological advice and in doing so ask the right questions. For his part the meteorologist is making progress not only by acquiring more records of wind but also by developing theoretical knowledge and applying new statistical techniques to his data. Consequently he is now in a better position to make reasonable estimates of probable maximum wind speeds not only for places for which satisfactory wind records are available locally but even for many where they are not.

The Meteorological Office receives frequent requests for information on probable maximum wind speeds not only for open sites on more or less level ground and for heights of up to about 100 feet, but also for city centres, hill and cliff tops and other places where natural topography or man-made obstacles may play an important part, and for heights up to 500 feet or more above the ground. They may relate to the design of all kinds of structures including radio and television masts, radar installations and radomes, radio telescopes, bridges and power stations, as well as buildings and their components.

The purposes of this paper are to indicate the type and quantity of wind data suitable for estimating design wind speeds that are available for the United Kingdom and to describe the procedures currently used in the Climatological Services branch of the Meteorological Office to provide appropriate information to inquirers. Special reference will be made to the problems of estimating maximum speeds at considerable heights above the ground and on unorthodox sites such as mountain tops and attention will be drawn to gaps in our knowledge that still have to be filled.

2. WIND DATA

The natural wind is usually turbulent, consisting of a succession of gusts and lulls with associated fluctuations in wind direction. These changes in speed and direction are very irregular and are made up of oscillations with periods ranging from a fraction of a second to many minutes, the eddies with which they are associated ranging in size from very small to very large. This can be deduced from an examination of the traces from a recording anemometer if the time scale of the chart is sufficiently open. The picture of wind structure thus obtained depends to some extent, however, on the sensitivity of instrument used and, as already indicated, on the time scale of the chart.

The standard wind-recording instrument used in this country for over 50 years was the Dines pressure-tube anemograph, and it is still in service at a large number of stations. A typical record is shown in *Fig.1*, which is a reproduction of part of the record from Tiree, Argyllshire, for 26th February, 1961, when a record gust of 101.5 knots (117 m.p.h.) occurred at 1750 G.M.T. The time scale of the chart, which is changed daily, is 15 mm to the hour. In recent years a different type of instrument, the M.O. electrical-cup anemograph, has been introduced and this is gradually replacing the pressure-tube instrument, mainly because the recorder can be placed at a considerable distance from the head, installation costs are less and servicing is easier. The sensitivity of the electrical anemograph is about the same as that of the Dines instrument, however, and although it has a strip-chart recorder, the time scale, as normally used, is 25 mm to the hour.

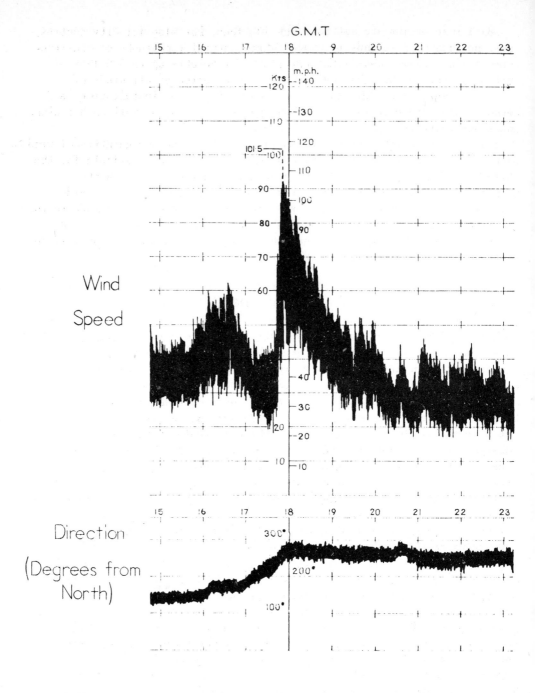

Fig.1. Record of wind speed and direction at Tiree from 1500 to 2300 GMT, February 26 1961.

Both instruments are sufficiently sensitive to provide accurate means over periods of 5 seconds if used with a chart with a sufficiently open-time scale. Strong gusts having durations of about 3 seconds or more are reliably recorded but shorter period gusts may not be fully registered.

The standard method of analysing these records is to tabulate the mean wind speed and direction over each hour and the highest gust for each day, plus some details of strong gusts, squalls etc. Thus the basic data that are available on extremes are the highest mean hourly values and the highest gusts, which can be taken to refer to periods of 3 to 5 seconds. The time scale of the standard charts and the labour involved make it impracticable to tabulate mean values over periods between these two. Thus if information is required about maximum speeds meaned over periods such as 10 minutes, 1 minute or 15 seconds these have to be estimated. I shall refer later to this question of the relationship between maximum wind speeds over various periods of time.

At present there are about 75 anemographs scattered over the United Kingdom from which wind records covering the 24 hours of every day are tabulated. This is a large number for a relatively small country although it is still not enough for some purposes. Summaries are published in the Monthly Weather Report and include inter alia, the maximum mean hourly wind speed and maximum gust recorded at each station. The records from these instruments, together with those from many others which have closed, constitute the basic data from which all our estimates of probable maximum speeds are made. Ideally all anemographs whose records are accepted by the Office must be set up on open level sites so that the wind recorded is that at a height of 10 metres (33 feet) above the ground at a place free from obstructions. In practice it is not always possible to do this and the instruments may have to be exposed at higher levels in order to avoid the effects of surrounding trees or buildings. Each anemograph is allotted an "effective height" (defined as the height at which it would register a speed equal to that actually observed, if there were no obstructions in the vicinity) based on a consideration of the various obstructions within 200-300 yards of the instrument.

3. STANDARDISATION OF EXTREME WIND SPEEDS

(a) Because the maximum speeds recorded by an anemograph depend both on its height above the ground (strictly on its effective height) and on the number of years of record, it is first desirable that all wind extremes should be reduced to the same standards, viz. an effective height of 33 feet above the ground and a return period of 50 or 100 years. The reduction to standard height may be made by applying formulae for the variation of wind

speed with height, established empirically; the reduction to a common return period by applying the statistical theory of extreme values to the annual extremes. A speed that has a return period of 50 years is that speed which on the average will be exceeded only once in 50 years; the chance of it being exceeded in any given year is 2%.

(b) Reduction to standard height

It is well established that the variation of mean wind speed with height under conditions of strong wind and over the first few hundreds of feet can most simply be represented by a power-law expression:-

$$\frac{v_h}{v_{10}} = \left(\frac{h}{10}\right)^{\alpha} \tag{1}$$

where v_h and v_{10} are the mean speeds at h metres and at the standard height of 10 metres above the ground, respectively, and α varies from about 0.1 to 0.4, depending on the nature of the terrain. A value of 0.17 has been adopted for general use in this country although it is most appropriate to level or gently rolling countryside, with only scattered trees or other obstructions, and, of course, to near-neutral stability, (Carruthers[1], Frost[2]). As the majority of our anemographs have effective heights which are within 50 feet of the standard height, the use of equation (1) with $\alpha = 0.17$ regardless of the nature of their surroundings, for reducing their maximum speeds to the standard height, should not introduce any appreciable errors. Whether the same formula can be used generally for estimating maximum speeds at heights of several hundreds of feet is a different matter and this problem will be considered later.

The variation of gust speed with height is less than that of mean speed; in other words gustiness decreases with increasing height. Measurements by Deacon[3] at heights up to 500 feet in Australia showed that maximum gust speeds vary with height in accordance with a power law expression with $\alpha = 0.085$. This provides values which are in very close agreement with those given by more complicated formulae due to Carruthers[1] and Bilham[4], both based on observations made in the United Kingdom. Deacon's formula is therefore used for reducing maximum gust speeds to the standard height. Incidentally, Deacon's observations of mean wind speeds fit a power law expression with an index of 0.16, the surrounding terrain being gently rolling grazing land with few trees.

(c) Application of statistical theory of extremes

The application of the statistical theory of extreme values (Gumbel[5]) to a series of annual extreme wind speeds permits the average number of years between recurrences of any particular speed or, alternatively, the speed which is likely to be exceeded only once in any given number of years, to be stated. The "predicted" extreme so obtained depends not on the highest single value on record but on all the annual maxima measured over a period of years. Maximum speeds obtained in this way also provide a more satisfactory basis for estimating design wind speeds because they permit the anticipated lifetime of the structure to be taken into account. In order to apply the statistical theory satisfactorily it is essential to have at least 10 and preferably at least 20 years' data. It is also necessary that the data should fit the theoretical distribution reasonably well. This has been found generally to be the case with wind records for the United Kingdom and the same thing has been found for the U.S.A. (Court[6], Thom.[7]).

Table I and fig.2, taken from Shellard[8], illustrate the procedure. The extreme values for each year are arranged in order of magnitude. Each is allotted a plotting position $p = \frac{m}{n+1}$ where m is its rank and n is the number of observations (years of record).

Table I
Annual maximum gust speeds at Cardington, 1932-54

Rank m	Highest Gust x mph	Year	Plotting position $p = \frac{m}{n+1}$	Reduced variate $y = -\log_e(-\log_e p)$
1	55	1953	0.042	-1.16
2	59	1950	0.083	-0.91
3	60	1941	0.125	-0.73
4	61	1951	0.167	-0.58
5	62	1952	0.208	-0.45
6	63	1937	0.250	-0.33
7	63	1939	0.292	-0.21
8	64	1942	0.333	-0.09
9	65	1933	0.375	0.02
10	67	1949	0.417	0.13
11	68	1948	0.458	0.25
12	69	1945	0.500	0.37
13	71	1940	0.542	0.49
14	72	1934	0.583	0.62
15	72	1944	0.625	0.75

Rank	Highest gust	Year	Plotting position	Reduced variate
16	76	1954	0.667	0.90
17	78	1943	0.708	1.06
18	78	1946	0.750	1.25
19	81	1932	0.792	1.46
20	82	1936	0.833	1.70
21	86	1938	0.875	2.01
22	88	1935	0.917	2.44
23	93	1947	0.958	3.15

The values are then plotted on a special probability paper on which the ordinate is the observed variable (wind speed) and the abscissa is p. Alternatively, values of $y = -\log_e(-\log_e p)$ may be computed or taken from published tables[9], in which case ordinary linear graph paper may be used, wind speeds being plotted against values of y. In *fig.2* fitted straight lines have been drawn in for both maximum gusts and maximum mean hourly speeds. The extreme speeds to be expected once in 50 or 100 years can be obtained by proceeding vertically from the 50 or 100 year mark on the return period scale to the intersection with the appropriate fitted straight line and thence horizontally to the ordinate where the required speed is read off.

(d) Results

Figs. 3 and 4 taken from Shellard[9] are maps of the distribution over the United Kingdom of the maximum mean hourly and maximum gust speeds, respectively, that are likely to be exceeded once in 50 years at 33 feet above the ground. They are based on records from 56 stations covering periods ending 1959 and were obtained as indicated in the two preceding sections. These maps give only a broad picture being derived from records from open and generally level sites at altitudes below 500 feet above sea level. However, for such sites it can be deduced that:-

(i) hourly mean wind speeds may exceed 70 m.p.h. on exposed western coasts with extreme gusts of 110 m.p.h. or more,

(ii) on exposed eastern coasts the highest hourly speed is about 60 m.p.h. with gusts of about 100 m.p.h.,

(iii) inland, maximum speeds are generally lower, with mean hourly speeds below 50 m.p.h. and gusts below 90 m.p.h. well inland in the Midlands and southern England,

(iv) in extensive built-up areas maximum mean hourly speeds tend to be further reduced but maximum gusts may be of practically the same magnitude as those over surrounding open country.

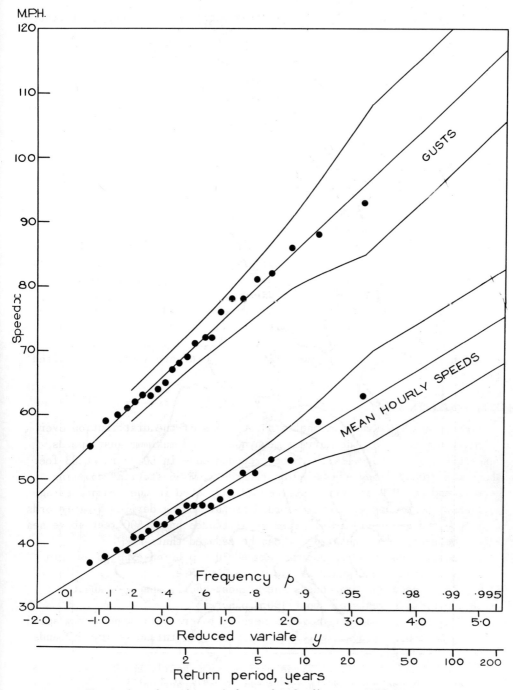

Fig. 2. Annual maximum wind speeds, Cardington, 1932-54

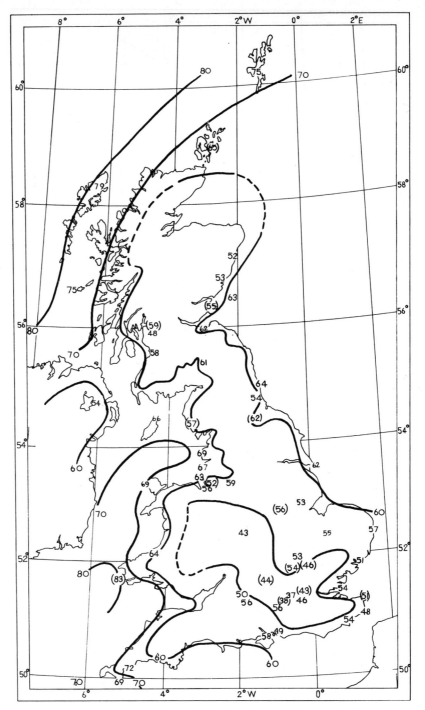

Fig. 3. Highest mean hourly wind speed (m.p.h.) at 33 feet likely to be exceeded only once in 50 years.
Values based on less than 15 years of record are bracketed.

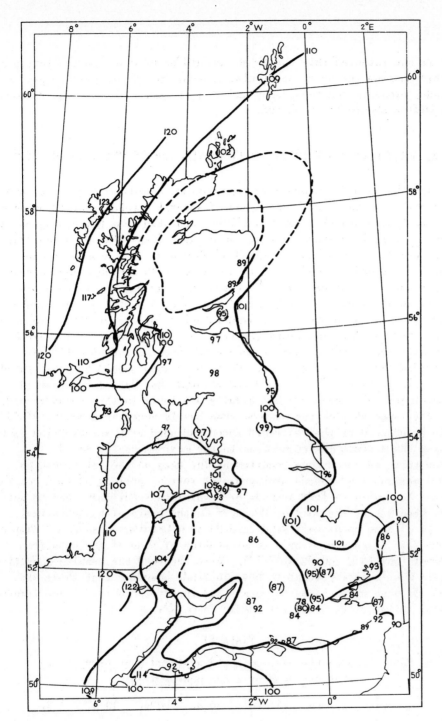

Fig. 4. Highest gust speed (m.p.h.) at 33 feet likely to be exceeded only once in 50 years.
Values based on less than 15 years of record are bracketed.

It is not intended that these maps should be used for interpolation of probable maximum speeds at particular places. Due regard must be paid to altitude, exposure and topography and it is recommended that the Meteorological Office should be consulted.

4. ESTIMATION OF MEAN SPEEDS OVER SHORT PERIODS OF TIME

So far the data discussed have been maximum speeds averaged over one hour and maximum gusts with durations of several seconds, mainly because these are the values that can readily be tabulated from British anemographs. In the American literature maximum speeds are often related to the "fastest mile of wind". On this basis a speed of 60 m.p.h. is a mean over one minute and a speed of 100 m.p.h. is a mean over 36 seconds. The current British Code of Practice (1952)[10] states that the maximum value of the mean velocity at an effective height of 40 feet over a period of one minute should be taken to be the normal wind velocity for deciding the exposure grading. Requests are sometimes received for advice on probable maximum speeds over periods even shorter than this. Scruton and Gimpel[11] have recommended that an averaging period of 15 seconds should be substituted for one minute in the British code. Realising that data for either 15-second or 1-minute means are not usually available in this country, however, they suggest that the maximum mean hourly speed at a height of 33 feet should be adopted as the reference speed and used with a conversion factor relating the maximum 15-second mean to the maximum hourly speed.

Thus there is an obvious requirement for data on the relationships between maximum wind speeds averaged over various periods of time ranging from one hour down to five seconds or less. This requirement has to some extent been met by the work of Durst.[12] On the basis of a statistical analysis of some open-scale wind records from Cardington and, for shorter periods than 5 seconds, from a consideration of some of the data from Ann Arbor (Sherlock and Stout[13,14]), Durst obtained the results summarised in *Table IIa*. It is important to bear in mind, however, that these are average figures which apply to open sites in level country. Other figures relating to the problem are set out in *Table IIb*.

Table II

(a) Ratio of probable maximum speed averaged over time 't' to that averaged over one hour (Durst)

t	1 hour	10 min.	1 min.	30 s.	20 s.	10 s.	5 s.	½ s.
Ratio	1.00	1.06	1.24	1.33	1.36	1.43	1.47	1.59

(b) Ratios between maximum speeds over various periods

	Ratio	Type of terrain	Source reference
10 s. means to 5 min. means	1.50	open country	15
5 s. " " 10 min. " (approx)	1.44	" "	3
3-5 s. " (max. gusts) to hourly means	(1.45-1.60 open exposures) (near coast) (1.60-1.80 " ") (inland) (2.00-2.20 city centres)		9
3-5 s. " " " 10 min. means	1.48	open exposure near coast	

The last line of *Table IIb* gives the result of a recent analysis carried out in the Meteorological Office of one winter month's records from Prestwick airport, using all occasions (404) when the 10-minute mean speed was equal to or greater than 20 knots. The distribution of the individual ratios of max. gust to mean 10 minute speed was as follows:-

Ratio	1.3	1.4	1.5	1.6	1.7	1.8	1.9
No. of occasions	23	131	177	50	20	1	2

It may be reasonably concluded from *Table IIb* that Durst's figures may be a little on the low side even for open country exposures, they will almost certainly be too low for urban and city exposures. Until further data become available, the following factors for converting maximum mean hourly speeds to maximum minute, 30 s. and 10 s. means are tentatively suggested as being suitable for routine use:-

	1 min.	30 s.	10 s.
(a) Open rural exposures	1.25	1.33	1.45
(b) urban and city exposures	1.45	1.60	1.80

Thus the "rule of thumb" for obtaining the maximum one minute speed, as required by the current British Code of Practice, is to add to the maximum mean hourly speed 25% for open country exposures and 45% for city exposures.

5. ALLOWANCE FOR TOPOGRAPHICAL EFFECTS

(a) Making suitable allowances for topographical effects when estimating probable maximum wind speeds at a particular place is not an easy matter. It usually consists of applying approximate rules or surmises which are based on general meteorological experience rather than on directly applicable

observational evidence. Some examples are given below.

(b) On well-exposed flat coasts the ratio of the maximum gust to the maximum mean hourly speed is about 1.5, inland it rises to about 1.75 and as the maximum gust speed is not greatly reduced it follows that the maximum mean hourly speed is reduced by about 15% at a distance of about 10 miles or more from the coast.

(c) On hill-tops which rise rather abruptly from a more or less level plain, or from the sea, maximum speeds are not much different from those at the same height in the free air upwind. For example, if the maximum mean hourly and maximum gust speeds at 33 feet above the plain, on a once in 50 years basis, are 50 m.p.h. and 75 m.p.h. respectively then the corresponding maximum speeds 33 feet above a hill top which rises to 300 feet above the plain will be approximately 72 and 91 m.p.h. (using the 0.17 and 0.085 power law formulae for mean hourly and gust speeds respectively). Some observational evidence to support this is presented in Section VII.

(d) In valleys which are orientated along a direction from which strong winds blow, these winds may be even stronger due to a funnelling effect. It is difficult to allow quantitatively for this without making measurements but the existence of the phenomenon may be known to local inhabitants or may be apparent from the presence of wind-pruned trees.

(e) On lee-slopes and to the lee of hills wind speeds will often be lower due to the sheltering effect of the higher ground, but in such situations the possibility of lee-wave effects has to be borne in mind. There is now good reason to believe that many of the strongest surface winds that occur from a westerly direction to the east of the Pennines are accentuated by the presence of lee waves, e.g. the exceptional and damaging Sheffield gale of 16th February, 1962.[16]

(f) Trees and buildings act as wind-breaks so reducing the mean speed of strong winds. Thus in towns and in well-wooded areas mean hourly speeds are of the order of 10 m.p.h. lower than over the surrounding country. The slowing up of the shorter period means is not nearly so marked because they are not subject to the effects of the increased surface friction to the same extent: the effect of the rough surface is to reduce the mean speed but to increase the gustiness, as indicated, for example, by the width of the trace on an anemograph chart such as *fig.1*. The arrangement of buildings in a town may cause channelling effects similar to those in a valley and it is often found that the distribution of damage in a gale can

be related to the lay-out of neighbouring roads and buildings. It is also known that a row of houses across a strong wind will create a lee eddy in which the surface wind is light or even reversed in direction but that at and above roof level there may be a region of markedly increased gustiness which can cause damage to other buildings downwind.

6. ESTIMATION OF MAXIMUM WIND SPEEDS AT CONSIDERABLE HEIGHTS ABOVE GROUND

In section III(b) the use of power-law formulae for the reduction of recorded wind speeds to standard height was discussed. How far can these same formulae be used to estimate maximum wind speeds at heights of several hundreds of feet above the ground, i.e. for estimating design wind speeds for structures with heights of 500 or even 1000 feet?

As mentioned earlier, power law formulae expressing the variation of mean wind speed with height have been established empirically by numerous workers in different parts of the world. The indices obtained for strong winds range from about 0.10 to 0.40 or higher, depending on the nature of the terrain. A good summary has been given by Davenport.[17]

The current British Code of Practice (1952) uses a power law index of 0.13 for heights up to 200 feet. A recent American source[18] recommends one seventh (0.143) for inland areas and heights up to 1000 feet, above which the speed should be assumed constant. For coastal areas, mainly to take account of hurricane conditions, it recommends 0.3 up to 600 feet where the basic "fastest mile" wind speed is 60 m.p.h. or less, or an index varying from 0.3 to 0.143 where the basic speed exceeds 60 m.p.h.; above 600 feet the speed is assumed constant. Observations made in this country suggest that an index of 0.17 is most appropriate, at least for open country, and that the increase with height continues up to at least 1000 feet.

The height at which the increase in mean wind speed becomes negligible, i.e. the height at which the effect of ground friction ceases, is that at which the pressure gradient is balanced against the centrifugal forces due to the earth's rotation and the curvature of the wind's path. This height, Z_G, depends, inter alia, on the average roughness of the underlying surface and probably varies over the range 1000 to 2000 feet, Davenport[17,19] has attempted on the basis of available experimental records to allocate suitable values of the power law index α and of Z_G to various types of terrain, and, by applying these values to extreme wind-speed data for the U.K. (taken from reference 8) he has deduced values for the parameters of the distribution of extreme gradient wind speeds over the country. He proposes that these should first be used to obtain the maximum gradient wind velocity

V_G at any place for any desired return period. Then, using the power law formula

$$V_Z = V_G \left(\frac{Z}{Z_G}\right)^\alpha$$

with values of Z_G and α chosen to suit the type of terrain, V_Z the probable maximum wind speed at height Z, for the same return period, can be calculated. The idea is an attractive one but it suffers from a number of practical disadvantages.

In the first place the parameters of the V_G distribution depend on the values of α and Z_G that were originally chosen, in a purely subjective manner, to fit the presumed roughness of the terrain surrounding each anemograph. Secondly, application of the method leads to results which are not consistent with actual measurements, as will now be demonstrated.

It has been established that a value of $\alpha = 0.17$ applies very well to the variation of speed with height up to at least 1000 feet at Cardington. The once in 50 years maximum mean hourly speed at 33 feet above the ground at Cardington, based on 28 years of record, is 53 m.p.h. Making the reasonable assumption that the gradient wind velocity is attained at a height of 500 m (1640 feet) we obtain $V_G = 103$ m.p.h. Davenports' own work (*fig.4* of reference 19) gives

$$V_G = 74 + 7.4 \log_e 50$$
$$= 103 \text{ m.p.h. over S.E. England}$$

The fact that the once in 50 years maximum gust speed at Cardington is 90 m.p.h. also indicates that the corresponding maximum gradient speed must be at the very least 90 m.p.h. A good starting point then is to assume that over S.E. England the maximum gradient wind speed exceeded only once in 50 years is 103 m.p.h.

Fig.5 shows wind profiles for $V_G = 103$ m.p.h. and for values of α ranging from 0.13 to 0.40, (a) assuming $Z_G = 500$ m (1640 feet) and (b) assuming that Z_G depends on α in the sort of way indicated by Davenport.

It will be noted that the ranges of speeds at the standard height of 33 feet resulting from these calculations and corresponding to a change in α from 0.13 to 0.40 are very large, amounting to 40½ and 47 m.p.h. in *figs.5(a)* and *5(b)* respectively. It will also be seen that while the range is greater for a varying than for a constant Z_G, the effect of any error in the assumed value of Z_G is of much less consequence than an error in the assumed value of α. Particular attention is drawn to the fact that an assumed value of $\alpha = 0.40$, suggested by Davenport as appropriate to city centres, leads to speeds of only just over 20 m.p.h. at the standard height. Measured speeds in the London area are about twice as strong as this.

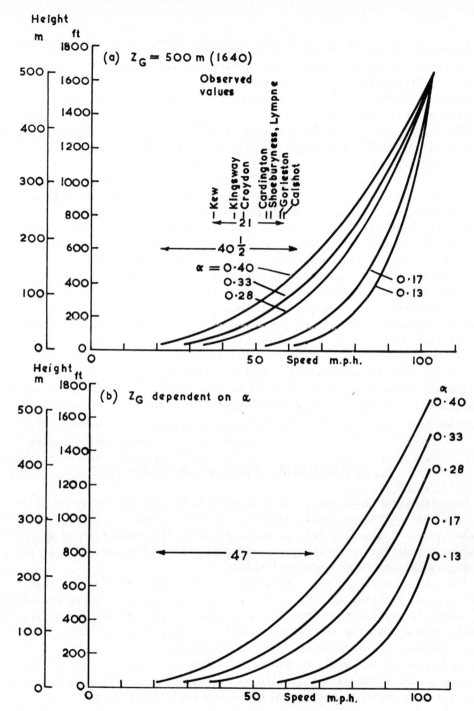

Fig. 5. Computed profiles of maximum mean hourly speed (once in 50 years) over S.E. England, as α and Z_G are varied, assuming $V_G = 103$ m.p.h; and comparison with observed values at 33 feet (10 m).

Measured values of the once in 50 year maximum mean hourly speeds at standard height in S.E. England are indicated, for comparison purposes, on *fig.5(a)*. Although the stations involved vary from openly exposed coastal stations to stations in built up areas, their maximum hourly speeds cover a range of only 21 m.p.h.

The comparison is brought out more clearly in *fig.6*. This confirms that a value of $\alpha = 0.17$ is appropriate for stations in S.E. England outside the London area. It also suggests that while an index of 0.40 would be quite inappropriate, a value of about 0.23 might be more suitable for metropolitan London than the 0.17 now in general use. Arrangements have been made for an anemograph to be set up on the new Museum Tower in London, at a height of about 550 feet above the ground, when the building is completed, probably in 1964. Data from this instrument in conjunction with those from low level stations should indicate definitely whether a value of $\alpha = 0.23$ would in fact be more appropriate. If maximum mean hourly speeds on windy days at 550 feet are about 90% stronger than those near the ground this will be the case, if they are only about 60% stronger retention of a general index of 0.17 will be justified. In this connection, it is suspected that some of the high values of α given in the literature for city centres may have arisen because the low level observations used may have been quite unrepresentative in that they related to sites sheltered by buildings rather than, as in London, to open sites, either a little above the general roof top level or in the centre of a large open space such as a park.

7. MAXIMUM WIND SPEEDS ON MOUNTAIN TOPS

Many inquiries for maximum wind-speed data have in recent years referred to radio and television masts and radar installations. For obvious reasons these are often sited on hill or mountain tops. The estimation of probable maximum speeds for such sites presents difficulties because until recent years few reliable data were available.

Two sets of mountain top anemograph records have recently been examined and compared with the records from the nearest low-level anemograph stations. The mountain stations are Drum in North Wales, 2528 feet above m.s.l., from which records covering the period October to mid-December, 1956 and April-June 1957 were obtained by courtesy of the Metropolitan-Vickers Electrical Co. Ltd. and Lowther Hill, Scotland, 2372 feet above m.s.l., where a M.O. electrical anemograph was installed in July 1961. Another anemograph has just recently begun to operate on Great Dun Fell, Westmorland, 2780 feet above m.s.l.

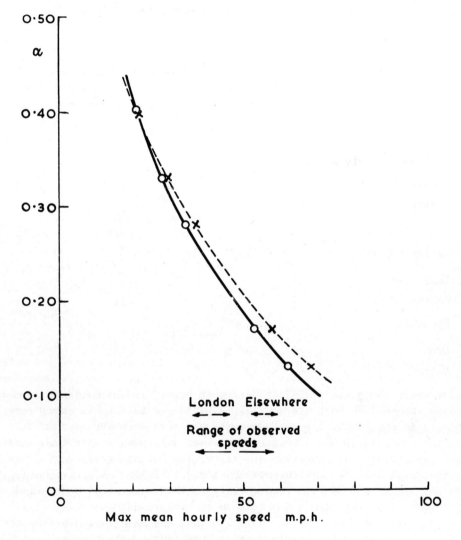

Fig.6. Relation between computed mean hourly speed at 33 ft (10 m) over S.E. England and assumed value of α
    (a) for $Z_G = 1640$ ft (500 m)   o———o
    (b) for Z_G dependent on α x----x
and comparison with observed values.

The maximum speeds recorded at Drum have been compared with those recorded at Valley, some 24 miles to the west-north-west and those recorded at Lowther Hill have been compared with those from Prestwick, some 34 miles to the west-north-west. The comparisons relate to windy days, defined for this purpose as days when the highest mean hourly speed at the mountain station exceeded 40 knots, thus giving about 50 such days for comparison at each pair of stations. The results are set out in *Table III*.

Table III

Comparison of maximum wind speeds on mountain tops with those at the nearest low-level anemograph stations on windy days

	Mean max. speed kt.	No. of days	Ratio of mean speeds
(a) Max. mean hourly speeds			
Lowther Hill (2412 ft)	47.6	51	1.82
Prestwick (68 ft)	26.2		
Drum (2548 ft)	50.7	45	1.89
Valley (76 ft)	26.8		
(b) Max. gusts			
Lowther Hill	69.2	51	1.58
Prestwick	43.9		
Drum	68.4	45	1.75
Valley	39.1		

The table shows that on windy days mean hourly speeds on exposed mountain tops at around 2500 feet are some 80-90% stronger than those at exposed places near sea level. The maximum gusts are also much stronger on the mountain tops, about 60% stronger at Lowther Hill than at Prestwick and about 75% stronger at Drum than at Valley. The heights given in the table are the heights of the instruments above m.s.l. Their "effective heights" were 30, 33, 20 and 38 feet respectively so that none of them departed appreciably from the standard height above the ground.

It may also be deduced from *Table III* that the mean ratios of the maximum gust to the maximum mean hourly speed at the four stations were 1.45, 1.67, 1.35 and 1.46 respectively, those from the mountain stations averaging 1.4 and being lower than those found at low-level stations even when these have very open coastal exposures.

It is of interest to compare the values in *Table III* with those that might have been predicted on the basis of the empirical power law formulae with indices of 0.17 and 0.085 for mean hourly and gust speeds respectively, assuming these to apply up to (a) 1000 feet and (b) 1500 feet with a constant speed above these heights, and also assuming that the maximum speeds on a mountain top are comparable with those in the free air at the same height. The results are shown in *Table IV*.

Table IV

Estimated mean maximum wind speeds (kt) at mountain stations, compared with observed values, assuming (a) Z_G = 1000 ft (b) Z_G = 1500 ft

	Estimated (a) Z_G = 1000'	(b) Z_G = 1500'	Observed
Max. mean hourly speeds			
Lowther Hill	46.9	50.0	47.6
Drum	48.0	51.1	50.7
Max. gusts			
Lowther Hill	58.9	60.6	69.2
Drum	52.4	53.9	68.4

It appears that a reasonably good estimate of the maximum mean hourly speed on an exposed mountain top can be obtained, when the only actual records available are those from a low-level station in the area, by using a power-law formula with index 0.17 up to about 1200 feet above which level speed is assumed constant. Use of the 0.085 power law formula for the estimation of maximum gusts at these heights may give a reasonable answer for free air gusts but it certainly underestimates those observed on mountain tops. It is suggested that these can best be predicted by applying a suitable gust factor to the predicted maximum mean hourly speed. An appropriate factor for 3-5 second gusts seems to be about 1.4. It also follows from this that the "add 25%" rule for estimating the maximum one minute speed will probably give too high a value for a mountain top and "add 20%" will probably provide a better estimate.

8. SUMMARY AND CONCLUDING REMARKS

In practice the estimation of maximum wind speeds for design purposes in the United Kingdom is thus based on statistically estimated maximum mean hourly and maximum gust (3-5 seconds) speeds likely to be exceeded once in 50 (or once in 100) years at 33 feet above the ground in open situations on more or less level ground. Estimates of maximum speeds over periods shorter than one hour, in particular the one minute maxima required by the B.S. Code of Practice, are based on Durst's analysis of open-scale records from flat unobstructed sites, but appropriate adjustments are made for sites, such as city centres and mountain tops, where the gustiness of the wind is known to depart considerably from that prevailing at sites similar to Cardington (ratio of maximum gust to maximum mean hourly speed outside the range 1.5 - 1.8). Variations of wind speed with height are based on power law formulae with indices of 0.17 and 0.085 respectively for mean hourly speeds and maximum gusts. Allowances are made for topographical effects where possible, as indicated in section V of the paper.

The possibility of basing estimates of maximum speeds on the maximum gradient wind speed in conjunction with suitable values of α and Z_G for the site concerned, as suggested by Davenport, has been considered. It is concluded that this cannot give satisfactory results unless much more reliable information becomes available about the values of α and Z_G that are appropriate to different types of terrain.

More information is required over cities concerning the variation of wind speed with height, and the relation between mean values over various periods. It is expected that some very useful data on these aspects will become available in a year or two when records should become available from an anemograph at 550 feet over central London. More information is also required about the effects of topography, in particular about the effect of lee waves on wind speeds near the ground and about the frequency with which atmospheric conditions leading to large-amplitude lee waves occur.

REFERENCES

1. CARRUTHERS, N. Variations in wind velocity near the ground. *Quart. J. roy. meteor. Soc.*, 1943, 69, 293.
2. FROST, R. The velocity profile in the lowest 400 ft. *meteor. Mag.* 1947, 76, 14.
3. DEACON, E.L. Gust variation with height up to 150 m. *Quart. J. roy. meteor. Soc.*, 1955, 81, 562.
4. BILHAM, E.G. Note on extreme gusts recorded in the British Isles, 1938, unpublished (but quoted in reference 1).
5. GUMBEL, E.J. Statistical theory of extreme values and some practical applications. *Appl. Maths. Ser., nat. Bur. Stand.*, 1954, 33.
6. COURT, A. Wind extremes as design factors. *J. Frank, Inst.*, 1953, 256, 39.
7. THOM, H.C.S. Distributions of extreme winds in the United States. *Proc. Amer. Soc. civ. Engrs.*, 1960, 86.

8. SHELLARD, H.C. Extreme wind speeds over Great Britain and Northern Ireland. *Meteor. Mag.* 1958, 87, 257.
9. SHELLARD, H.C. Extreme wind speeds over the United Kingdom for periods ending 1959. *Meteor. Mag.* 1962, 91, 39.
10. LONDON, BRITISH STANDARDS INSTITUTION. *British Standard Code of Practice, C.P. 3 - Ch.V. Loading,* 1952, 12.
11. SCRUTON, C. and GIMPEL, G. Memorandum on wind structure with reference to the wind pressure clauses of B.S. Code of Practice, C.P. 3 - Ch.V. *Nat. Phys. Lab.* NPL/Aero/361, 1958.
12. DURST, C.S. Wind speeds over short periods of time. *Meteor. Mag.* 1960, 89, 181.
13. SHERLOCK, R.H. and STOUT, M.B. Picturing the structure of the wind. *Civ. Engng. Easton, Pa,* 1932, 2, 358.
14. SHERLOCK, R.H. and STOUT, M.B. Wind structure in winter storms. *J. Aeron, Sci.,* 1937, 5, 53.
15. SHERLOCK, R.H. Variation of wind velocity and gusts with height. *Proc. Amer. Soc. civ. Engrs.,* 1952, 78, 126.
16. AANENSEN, C.J.M. and SAWYER, J.S. "The gale of February 16, 1962 in the West Riding of Yorkshire". *Nature,* 1963, 197, 654.
17. DAVENPORT, A.G. Rationale for determining design wind velocities. *Proc. Amer. Soc. civ. Engrs., J. struct. Div.,* 1960, 86, 45.
18. Wind forces on structures, final report of the Task Committee on Wind Forces of the Committee on Loads and Stresses. *Proc. Amer. Soc. civ. Engrs., J. struct. Div.,* 1961, 126 Part II, 1124.
19. DAVENPORT, A.G. The application of statistical concepts to the wind loading of structures. *Proc. Inst. civ. Eng.* 1961, 19, 449.

PAPER 2

THE RELATIONSHIP OF WIND STRUCTURE TO WIND LOADING

by

A. G. DAVENPORT
(University of Western Ontario)

THE RELATIONSHIP OF WIND STRUCTURE TO WIND LOADING

by

A. G. DAVENPORT
(University of Western Ontario)

SUMMARY

TO evaluate wind loads on structures a knowledge of wind structure on both a macro- and micrometeorological scale is necessary. The former is concerned with the overall climatic properties of the wind and the recurrence of extremely destructive winds: the latter is concerned with the details of the flow in the earth's boundary layer - the wind velocity profile and the gustiness. Both aspects are discussed.

In a neutrally stable atmosphere (appropriate to strong winds) properties of the mean flow are shown to be dependent almost entirely (except near mountains) on the roughness of the ground which in nature can vary radically between open water and heavily built up urban areas. Simple power law velocity profiles with appropriate exponents and having a common gradient velocity at the upper limit are found to give good approximations to observed and predicted profiles over surfaces of widely differing roughness. The transition in boundary-layer flow which occurs when the wind blows from one surface to another of different roughness is examined.

The statistical properties of turbulence in strong winds are described in terms of the spectra and scales as well as the probability distributions. It is suggested that reasonably reliable estimates of these quantities can be made from a knowledge of the mean wind velocity and the ground roughness. A "two-dimensional jet" model of boundary-layer flow in the wind tunnel suggested by Townsend is shown to correspond to many of the properties of turbulence in strong winds.

The problems relating to the establishment of an "extreme-mean-wind-velocity" field over a country are discussed. For this purpose it is suggested that the use of a gradient-wind-velocity field has many advantages to offer. A map of the extreme-hourly-gradient-wind parameters over the United Kingdom is shown as an illustration. From this, surface wind velocities (and hence the gust spectra) can be found if the ground roughness can be estimated. The advantages of site-tests for determining wind properties are pointed out.

1. INTRODUCTION

Although wind loads are a major factor in the design of many structures, the nature of the wind itself is not a subject with which structural engineers are generally familiar. This state of affairs is no doubt partly due to the interdisciplinary nature of the subject, and partly because of the lack of emphasis usually given to it in structural engineering curricula. As a result however, structural design has tended to become compartmentalized: the estimation of wind-loads is often divorced from the study of the structure itself and the responsibility for their estimation delegated to others. Indeed, to a few structural engineers destructive winds may be little more than unpredictable acts of fate capable of little or no scientific description!

This somewhat defeatist approach may explain why apparently so little incentive has been given to research in the meteorological areas related to wind loading while in more clearly defined areas - such as the estimation of pressure distribution on structures from wind-tunnel models - much more effort has been expended. Next to no research, for example, has been carried out into the wind structure over an urban area in spite of its obvious relevance. Only recently has the climatology of high winds been examined. Subjects such as the influence of terrain roughness and topography on the wind properties have been almost untouched (with the exception of one or two notable contributions such as that due to Jensen) and certainly have had no impact on wind loads.

The intention in this paper is to try to point out the possible direction a scientific formulation of the wind properties relevant to structural design might take. This cannot be done, however without first establishing those features of structural behaviour which are liable to have a bearing on the subject and which define the terms of reference.

1.1 Wind-load criteria

The most serious consequence of wind action, the collapse of the structure, can arise either through a single application of a stress greater than the structural material can sustain at a sufficient number of points to render the structure a mechanism, or it can arise through repeated applications of somewhat lower stress levels which gradually lead to the fatigue failure of the members. Both types of collapse have occurred: the Tay Bridge disaster of 1879 is only one of several examples amongst large bridges of the former type of failure due to wind: examples of fatigue failure have been mainly amongst flexible structures such as street lamp standards and the hangers of suspended bridges as in the instance of an arch bridge at Tacony-Palmyra in 1929.

There are other consequences of wind loading which also must be considered, such as large deflections or accelerations. Although not a direct criterion of structural failure these can often sufficiently impair the utility of a structure, to be a primary design consideration. One can find examples of this in the excessive deflection of directional radio antennae, the clashing of power line conductors, and in the feelings of insecurity that large deflections and accelerations produce on occupants of tall buildings and bridges.

All these effects of the wind have this much in common; they all depend not only on the magnitude of the load but also on the time sequence of its application. This last fact which is a consequence of the dynamic characteristics of structures, is not widely recognized by structural engineers (or meteorologists) although its significance is greatly increasing as discussed in another paper.

A further factor which must be considered and which is directly related to the economic utility of a structure is its anticipated useful lifetime. It is seldom that this can be stated explicitly (except in the rather artificial sense used by accountants in writing off a structure in their books) although it is very much a fact of life that some structures are intended to last much longer than others. A cathedral is expected to last longer than a stop-gap "prefab" house or a radio mast in which electronic obsolescence is likely to have set in within 20 or 30 years. It is clearly economically desirable to reflect these differences explicitly in structural design and correspondingly in the wind loading.

These factors define the terms of reference in formulating design wind loads. It emerges that what is required is a statement defining first the magnitude of the wind speed at the site of the structure, second its time dependence both on a time scale characteristic of its lifetime and also of its dynamic response, and third its spatial characteristics in the region of aerodynamic influence surrounding the structure.

2. THE STRUCTURE OF THE WIND

2.1 General

The source of the wind's energy can of course be traced back to the sun. In very simple terms solar radiation, which is far more intense at the equator than the poles, tends to cause differential heating of the earth's surface which in turn gives rise to gradients of pressure in the atmosphere. These gradients of pressure can be inferred from the contours of equal barometric pressure (isobars) shown on weather maps such as that shown in Fig.1. At heights greater than 1000 ft or so, outside the influence of the frictional forces near the ground, the wind attains what is known as the gradient velocity. Its magnitude is a function of the latitude, the curvature of the isobars and their spacing i.e. pressure gradient.

Closer to the ground the airflow is slowed down by the drag forces at the surface and which are transmitted upwards by the virtual Reynolds stresses which result from the momentum exchange between layers due to turbulence. The rougher the ground the greater the drag force, the Reynolds stresses, the turbulent intensity, the retardation at the surface and the gradient height.

It follows that the wind at a point near the earth's surface will be characterized first by the large-scale movements of the pressure systems giving rise to the gradient wind and second by the modifying influence of the ground surface. It turns out that the two processes give rise to variations in wind speed having completely different time scales. This is indicated in Fig. 2 in which the spectrum of horizontal wind speed near the ground is shown for an extended range of frequencies.

Such a spectrum is analogous to the spectrum of light formed by passing light through a prism and which expresses the distribution of light energy with wavelength. One method of obtaining a wind-speed spectrum is to convert the wind-speed signal into its electrical analogue pass this through a range of filters having different characteristic frequencies and measure the output energy on a watt meter. In Fig. 2 the spectrum is plotted on a logarithmic frequency scale in such a way that the area under the curve between any two frequencies is proportional to the energy within that frequency range.

The spectrum is seen to be characterized by several prominent peaks, with several intervening ranges of low amplitude. On the right, the high frequency end, there is a broad peak with a maximum at a period of between 1 and 2 minutes per cycle. This part of the curve (which was in fact evaluated from records made in hurricane Carol) is characteristic of turbulence generated almost entirely mechanically by shear stresses at the ground surface: in high winds convection plays a minor role not only because the air is so churned up that thermal instabilities do not get a chance to build up; but also because the shear stresses and mechanical turbulence build up in proportion to the square of the wind speed and very soon swamp any contribution that convection might otherwise make even if conditions were favourable.

Between periods of 5 minutes and about 4 - 5 hours the spectrum contains very little energy implying that there is very little variation in wind speed occurring at these rates. At a twelve hour period there is a minor peak corresponding to the lulls in wind speed which generally occur at sunrise and sunset. At a period of about four days there is a major peak corresponding to the movement of large-scale weather systems such as those in Fig. 1: everyday experience tends to confirm that weather conditions fluctuate with a period of this order.

Spectral estimates for periods greater than about two months were not made by Van der Hoven but experience indicates that another very predominant peak occurs at a period of a year. This corresponds to seasonal fluctuations in wind speed which come about from the greater average difference in temperature between the equator and poles (which is the effective operating range of the earth acting as a heat engine) during winter than in summer. Other peaks might in fact be present for diurnal and lunar cycles.

Beyond a period of one year the shape of the spectrum is speculative. However it is difficult to conceive any other physical process capable of causing longer period fluctuations other than the eleven year sun spot cycle which can only be a weak effect.

The concept of atmospheric motion as compounded of superimposed trains of wind waves of different frequencies and of amplitudes dictated by the spectrum is a useful one to have. An extremely strong wind can then be visualized as the simultaneous arrival of the peaks of several important wave trains. If the probability distribution, defining the proportion of the total time the wind speed exceeds a given value, were known - as well as the spectrum - it would theoretically be possible to estimate such quantities as the time interval between recurrences of extremely high wind speeds, which in part at least is the object of our inquiry. This approach does not however turn out to be very fruitful mainly because of the difficulty of estimating the left hand, low frequency portion of the spectrum; also the same results can be derived more directly as will be explained. The immediate usefulness of this approach lies in its aid to understanding of the processes involved.

It is important to realize that although a spectrum such as Fig. 2 will only apply to a particular site and to a particular height above ground, the general form of the spectrum and the position of the peaks may nevertheless be expected to remain very much the same regardless of the geographical locality, the nature of the terrain and the height above ground. One of the most important distinctions that it appears can be made is between the fluctuations of a macrometeorological kind such as the movement of large-scale pressure systems, seasonal variations etc. and those which are of a local, micrometeorological kind and associated with the flow characteristics of the boundary layer itself. In more simple parlance these two categories of fluctuation might be described as "weather-map fluctuations" and "gusts."

From Fig. 2 it appears that these two types of fluctuations are separated by a gap extending from roughly five minutes to five hours. This gap is important to our evaluation of wind loads for several reasons. It enables a clear cut distinction to be made between gusts and weather-map disturbances and furthermore their causes: in another sense this distinction can be regarded as between gusts and the mean wind where the mean wind is characterized by the average velocity over some period within the spectral gap. Fortunately almost all routine meteorological measurements of wind speed - with the exception of so-called peak gusts - are averaged over periods within the spectural gap. Mean hourly wind speeds are for example recorded in the U.K. and Canada, mean five-minute speeds in the U.S.A., mean ten-minute speeds in Japan etc. Because of the low amplitudes in the spectral gap all these various average speeds are in high winds almost equal. Thus it would seem that the meteorological measurements made over many years have been referred to averaging periods which (by good luck or good judgment) happen to be most suitable for making long term estimates of wind loads. From the experimental point of view the spectral gap enables stable estimates of the gust spectrum to be made from records of fairly short duration - 20 minutes or so. Because structures have natural periods of vibration of less than 1 minute it turns out that the part of the spectrum relating to gusts has high importance in the evaluation of wind loading.

The existence of this gap although not proved incontrovertibly, now seems strongly probable, certainly in high winds when, as explained earlier, lower frequency convective turbulence is relatively speaking far less significant. Van der Hoven demonstrated that it existed in a wide variety of weather conditions and localities and the writer has confirmed its existence in a run of very high wind measured at the Severn River railway bridge in Gloucestershire (kindly loaned by Messrs. Freeman, Fox and Partners.[11]) It is attributed to the lack of any physical processes capable of generating fluctuations at gap frequencies.

If under some circumstances the gap does not exist and the "weather map," convective and other non-mechanical disturbances extend the macro-meteorological range of the spectrum so that it overlaps and augments the gust spectrum, the task of evaluating the dynamic gust response will become problematic and the long-time meteorological records of hourly or five-minute-average winds of far less value. Fortunately however, this does not appear to be true, and, in the opinion of the writer, the model upon which wind loads are based should at present, assume the existence of this gap in the overwhelming majority of destructive winds.

The conditions under which this assumption might be in greatest error are intense local storms such as severe frontal squalls, thunderstorms and, in the extreme case, tornadoes. Although these types of storms are known to occur frequently in several regions of the world they are generally of such local extent, affecting only one or two square miles at a time, that the likelihood of a destructive storm of this type striking a particular site is negligibly small. This is evinced partly by experience in the notorious tornado belt of the U.S. midwest where it is never considered necessary to design against their contingency. It will be assumed in this discussion, that local storms of these types are either sufficiently rare or otherwise harmless that, ignoring the special design conditions they may present, will not affect the overall statistical picture of the wind loading arrived at from a study of large scale destructive storms for which the assumption of a spectral gap seems valid.

The model of the wind which emerges from this is one in which the overall flow characteristics (indicated by the gradient wind) are governed by large scale, slowly varying macrometeorological processes having little significant periodicity less than two or three hours. Nearer the surface this flow is modified by the boundary layer in which the mean flow is retarded and mechanical turbulence is generated.

Our model of the wind near the ground is thus analogous to that near the boundary layer of a wind tunnel in which the speed of the tunnel fan is slowly varied, at a rate which, (by analogy to the spectral gap) is much slower and of longer periodicity than any turbulent fluctuation. To take the analogy further the roughness of the wind-tunnel wall should be variable to correspond to the full-scale roughnesses of open water at one end of the scale and a built-up urban area on the other end. Under these conditions the mean velocity at a given height in the boundary layer will be proportional to the velocity at the tunnel centre line and will vary at the same rate, its magnitude will be a function of the roughness of the surface and hence the boundary-layer profile. The mean square of the fluctuation velocity in the boundary layer will depend on the surface shear stress which in turn depends on the surface roughness and the square of the mean velocity. The same general characteristics may be expected to pertain in high winds.

3. CHARACTERISTICS OF THE MEAN FLOW IN THE EARTH'S BOUNDARY LAYER IN HIGH WINDS

3.1 General

Above the layer of frictional influence near the surface the air moves purely under the influence of the pressure gradients and attains what is known as the gradient velocity. From the equation of motion for the air it is found[30] that the magnitude of the gradient wind round a low-pressure region (inducive of higher winds) is given by

$$V_G = R \omega \sin \lambda \left[\sqrt{\frac{\frac{dp}{dn}}{\rho R \omega^2 \sin^2 \lambda} + 1} - 1 \right] \qquad (1)$$

in which $\frac{dp}{dn}$ is the pressure gradient, ω is the angular velocity of the earth, λ is the latitude, ρ is the air density, and R is the radius of curvature of the isobars. The direction of the wind outside the region of frictional influence is parallel to the isobars as shown by the arrows in Fig. 1. Usually the fraction inside the square root sign is considerably less than unity and by binomial expansion we can write

$$V_G \approx \frac{\frac{dp}{dn}}{2\rho\omega\sin\lambda}\left\{1 - \frac{\left[\frac{\frac{dp}{dn}}{2\rho\omega\sin\lambda}\right]}{2R\omega\sin\lambda} + \right. \quad (2)$$

The second term in the expansion is only significant for small radii of curvature, and under most circumstances a good estimate is given by the first term

$$V_G \approx \frac{\frac{dp}{dn}}{2\rho\omega\sin\lambda} = \frac{\frac{dp}{dn}}{\rho f} \quad (3)$$

where $f = 2\omega\sin\lambda = 1.458 \times 10^{-4} \sin\lambda$ sec^{-1}.

This is known as the geostrophic wind and is that associated with the Coriolis acceleration due to the rotation of the earth.

The height at which the gradient velocity is attained will be denoted by Z_G and is generally of the order of 1000-2000 ft.

Closer to the ground the airflow is slowed down by the drag forces at the surface and the virtual Reynolds stresses which result from the momentum exchange between layers due to turbulence. In high winds the boundary layer is fully turbulent over all natural surfaces; the Reynolds stresses exceed the direct viscous stresses by several orders of magnitude and the latter may be neglected. Viscosity does, however, play an important role in controlling the rate of dissipation of the turbulence and hence indirectly the Reynolds stresses also. The rougher the ground, the greater the drag force at the surface, the turbulent intensity, the Reynolds stresses, the gradient height and the retardation at the surface.

The overall region of frictional influence below Z_G is termed the planetary boundary layer. It can be broken down into at least two sublayers, principally the surface boundary layer, in which the shearing stress is approximately constant, and a transition region in which the shearing stress falls off from the constant value of the surface layer to the practically zero value in the free atmosphere.

In the surface layer which extends up to roughly 200 ft ($\pm$ 100 ft) the wind velocity profile appears to be accurately defined by the Prandtl logarithmic profile. This is based on the assumption first of constant shearing stress and second of a so-called mixing length (analogous to the molecular mean free path) proportional to the height. It is valid in conditions of neutral stability (already pointed out to be appropriate for high winds) and is written

$$\frac{V_Z}{V_*} = \frac{1}{k} \ln \frac{Z}{Z_o} \qquad (4)$$

where V_Z is the mean velocity at height Z, $V_* = \sqrt{\frac{\tau_o}{\rho}}$ is the "friction velocity", Z_o the roughness length, k von Karman's constant (≈ 0.4), τ_o is the shear stress at the surface and ρ the air density. Related to the friction velocity is the surface drag coefficient κ such that

$$\tau_o = \kappa \rho \bar{V}^2 \qquad (5)$$

and

$$\kappa = \left(\frac{V_*}{\bar{V}}\right)^2 = \frac{k^2}{\ln^2 \frac{Z}{Z_o}} \qquad (6)$$

The logarithmic profile agrees well with experimental measurements in the wind-tunnel boundary layer and in the surface layer over natural surfaces of roughnesses varying between smooth mud flats, open water and thick grass. It has yet to be demonstrated experimentally to be characteristic of urban areas. Above the surface layer the effect of the Coriolis force increases, the effect of the surface roughness decreases and the profile departs significantly from the logarithmic form.

Several approaches have been taken to formulate the mean-wind profile through the deeper planetary boundary layer up to the height at which there is no systematic deviation from the wind motion implied by the barometric pressure gradient and geostrophic forces. Most of these have introduced the concept of the eddy viscosity. This is a means for defining the virtual shear stresses which arise from the exchange of momentum between layers. Precise formulations have generally floundered on the problem of defining the variation of the eddy viscosity with height. An assumption of constant eddy viscosity has led to the so-called Ekman spiral. Other models assuming power-law variations of the eddy viscosity with height have been put forward by Prandtl and Tollmien and Kohler[30]. A two-layer model which implies a mixing length which increases linearly with height to the top of the surface layer and above that decreases linearly to the top of the planetary layer, has been suggested by Rossby and Montgomery.[26]

These solutions account for several observed characteristics such as the systematic deviation of the wind direction from that of the isobars nearer the ground (greater over rougher surfaces). In spite of their greater sophistication, however, these expressions all contain quantities which can at present only be defined empirically. Their overall reliability in providing numerical predictions of the wind-speed profile do not appear to be greater than the simple power law profile given by

$$\left(\frac{\bar{V}_Z}{\bar{V}_1}\right) = \left(\frac{Z}{Z_1}\right)^\alpha \qquad (7)$$

in which $\bar{V}_1$ is a reference velocity at height Z_1 and α is a constant. Because of its simplicity and the lack of any expression yielding better accuracy, this seems a suitable profile for wind loading purposes.

If this curve is to represent the velocity throughout the boundary layer then at some height Z_G the velocity must attain the gradient value $\bar{V}_G$. In this case

$$\frac{\bar{V}_Z}{\bar{V}_G} = \left(\frac{Z}{Z_G}\right)^\alpha \qquad (8)$$

From this, if the index α and Z_G are known the wind velocity at any height in the boundary layer can be expressed as a ratio to the gradient wind. Both parameters, it seems, are principally functions of the roughness of the ground.

To determine values of α and Z_G for different natural surfaces the writer in an earlier paper[9] has surveyed the published measurements of wind profiles in high winds for some 19 different localities ranging in roughness from open water to the centre of a large city. The measurements varied in quality, some being based on a series of measurements and made over a considerable height range, in other instances they were based on isolated measurements or only over relatively shallow height ranges up to 100 ft. From these data and other information relating to the ratio of gradient to surface winds emerged a reasonably consistent picture from which the average values of α and Z_G given in Fig. 3 were derived. The profiles typifying three types of terrain - open terrain, wooded country and an urban centre - are shown in Figs. 4 and 5.

The main body of evidence for these curves was given in the previous paper and it is not the intention to repeat this here: however, since this was written other information has come to light which relates to these curves.

A study by Johnson[22] of the relation between the mean (hourly) wind velocity at 2 metres, 40 feet and 1000 feet at the Suffield Experiment Station near Medicine Hat, Alberta, has shown that $\frac{V_{40}}{V_{1000}} = 0.60$ and $\frac{V_{2m}}{V_{1000}} = 0.32$ in conditions of neutral stability. The site is in the heart of the so-called "short-grass country" in the unrelieved flatness of the Canadian prairies. If the 1000 foot velocity is taken as representative of the gradient wind these ratios provide confirmation of the "flat, open country" curve of Fig. 5. This curve also agrees well with Taylor's measurements in strong wind over Salisbury Plain (quoted by Sutton in "Micrometeorology") from which the same ratio was found to be 0.61.

A revealing study of the action of surface roughness on the wind was undertaken by Jensen in Denmark and given in his notable thesis on "Shelter Effect" (Jensen - 1954). Anemometers were set up at 2m height along two east-west lines running from coast to coast across the Danish peninsula (see Fig. 6). The length of the lines were 47½ miles (76 km) and 63 miles (101 km) and instruments were spaced at roughly 4 and 6 mile intervals on the two lines respectively. Simultaneous measurements of wind speed were made at heights of 2 metres on each of the lines (2 sets on line 1 and 3 in the case of line 2) on the occasions when the gradient wind was from the west at approximately 20 m/s (45 m.p.h.).

In Fig. 6 the velocities at each of the anemometers are given as a ratio to the gradient wind. Also shown is what Jensen terms the "roughness of great order" along the lines. This was calculated from the average frontal area of the obstructions per unit horizontal area of terrain multiplied by an "effect figure" which ranged between 2 and 4 for coniferous hedges and trees and between 1 and 3 for deciduous (according to spacing) and 4 for buildings, gardens and small plantations. These "effect figures" were determined from wind-tunnel measurements of the wind profile downstream from series of screens of varying porosity.

These curves in Fig. 6 indicate clearly the strong influence of surface roughness on the wind strength and tend to confirm the curves in Figs. 4 and 5.

The effect of surface roughness on two further parameters, the roughness length Z_o and the surface drag coefficient is also important (see Eqs. 4 and 6).

Values for Z_o are usually estimated directly from the logarithmic wind profile near the ground. Representative values for roughnesses up to thick grass are quoted by Sutton in "Micrometeorology": for rougher surfaces we have to turn elsewhere. Measurements over wooded areas have been carried out at Brookhaven Laboratory and at Munich. At the former the accepted value for Z_o is 1 metre[14] - while Lettau[21] quotes values of 20 cm from the profiles obtained by Baumgartner[2]. An analysis by Davenport[10] of some somewhat sparse data obtained by Jensen, and Shiotani over urban areas indicate that here the roughness length is of the order 1-3 metres.

Values of the drag coefficient have been measured in four ways:- from observations on the approach to the geostrophic wind, from wind profiles near the ground, by direct measurement using drag plates, by measurement of the Reynold's stress near the ground (mean product of longitudinal and vertical fluctuation velocities). For open sea Sutton[30] quotes Sutcliffe's value of $\kappa = 0.0005$ (with 10 metre reference height): for rough grassland 0.005 seems an accepted value: for a wooded area such as Brookhaven the roughness length 1m indicates $\kappa \approx 0.030$ while the Reynolds stress indicates $\kappa = 0.015$, which agrees with the value obtained from the wind profile for the Munich forest data mentioned above. If a roughness length of 3m is accepted for a city then the coefficient of drag turns out to be about 0.050.

These surface drag coefficients are, in fact, close to those encountered in artificial surfaces such as pipes, aircraft wings, and ships plating. It is also practice to represent the boundary-layer velocity profile over these surfaces by a power law (such as the well known 1/7 power law). Nunner[23] has shown (see Fig. 7) from pipe flow experiments that there is a systematic relationship between the index of the power law and the surface drag coefficient, as defined by $\left(\dfrac{V_*}{V_{AV}}\right)^2$ (see equation (2)), which can be expressed empirically by

$$\alpha = \sqrt{8\left(\dfrac{V_*}{V_{AV}}\right)^2}$$

where V_{AV} denotes the average velocity in the pipe. Turning to the profiles derived from the natural wind - the parameters of which are given in Fig. 3 - it can be shown that the average velocity in the boundary up to the gradient height can be expressed in terms of the 10 metre reference velocity as

$$\dfrac{V_{AV}}{V_{10}} = \dfrac{1}{1+\alpha}\left(\dfrac{Z_G}{10}\right)^\alpha = \emptyset \text{ (say)}$$

and that

$$\left(\dfrac{V_*}{V_{AV}}\right)^2 = \dfrac{\kappa}{\emptyset^2}$$

Three typical values of this function are also plotted in Fig. 7 as a function of the corresponding power law exponent and are seen to be well within the scatter of the experimental results. This interesting result indicates the universality of the turbulent boundary layer properties on two vastly different scales and for two different media.

The main objections to the above power law approach to defining the mean wind velocity profile are its empiricism and its dependence on the non-fundamental parameters α and Z_G both of which have a somewhat nebulous physical reality. The justifications for adopting the approach are first simplicity and second the fact that both α and Z_G or functions of these quantities can, it seems, be systematically related to the fundamental parameters defining the roughness of a surface or otherwise its effects, principally Z_o and V_*. A more direct approach is to group together the various parameters which do have some physical reality using dimensional arguments: two parameters of significance emerge

$$\frac{V_G}{fZ_o} \text{ and } \frac{V_*}{V_G} \left(\text{or } \frac{V_*}{fZ_o} \right)$$

All of these quantities are directly measurable (in contrast to α and Z_G which can only be inferred indirectly). V_G, the geostrophic wind, is directly translatable into the pressure gradient by the transformation

$$V_G = \frac{\frac{dp}{dn}}{\rho f} \quad \text{(see equation 3)} \tag{9}$$

where $f = 2\omega \sin \lambda$ (the ingredient of the Coriolis acceleration due to the earth's rotation): V_* can be directly translated into the shear stress at the surface: Z_o is the roughness length and constitutes a statistical measure of the average size, shape and arrangement of the surface obstructions.

The first of these parameters $\frac{V_G}{fZ_o}$ has been termed the Rossby number Ro. The second $\frac{V_*}{V_G}$ is by nature of a drag coefficient (or the square root of the more conventional form of drag coefficient such as κ) and Lettau[21] has termed it the geostrophic drag coefficient C_a. (The subscript denotes an adiabatic lapse rate i.e. neutral stability.)

As outlined, a number of theoretical models of the wind in the boundary layer have been suggested[26,30]. All are based on inherent assumptions (concerning the variation of eddy viscosity, mixing length etc. with height) which can explicitly or implicitly be written in the form

$$\frac{V_G}{fZ_o} = Ro = F_1 \left(\frac{V_*}{fZ_o} \right)$$

or alternatively

$$C_a = \frac{V_*}{V_G} = F_2 (Ro) \tag{10}$$

Taylor[32] has shown that, in spite of the range of inherent assumptions, several of the theoretical solutions can be matched extremely closely by simple power law relationships of the type

$$\frac{V_G}{fZ_o} \propto \left(\frac{V_*}{fZ_o} \right)^m$$

which is directly equivalent to

$$C_a \propto Ro^p \text{ where } p = \frac{1-m}{m} \qquad (11)$$

(Taylor ascribes this similarity to the dominance of the boundary conditions on the result viz. - constant shear stress near the surface and zero shear stress as the gradient velocity is approached). The same boundary conditions are assumed in all models.

In the several practical instances and theoretical models Taylor found p to be approximately -0.09. The validity of this expression tends also to be confirmed by some data collected by Lettau[21] for localities of varying latitude and roughnesses ranging between open sea and forest. These are shown on a log. plot of C_a versus Ro in Fig. 8 and compared with the line for p = -0.09. The full expression for the line is

$$C_a = 0.16 \, Ro^{-0.09} \qquad (12)$$

To find the velocity of the wind at a height Z near the surface in terms of the gradient wind, it is now necessary only to multiply both sides of the logarithmic profile (equation 4) by $\frac{V_*}{V_G}$ yielding

$$\frac{V_Z}{V_G} = \frac{C_a}{k} \ln \frac{Z}{Z_o} \qquad (13)$$

This expression is evaluated in Fig. 9 for heights of 10 m and 2 m as a function of the roughness length Z_o for the practical range of wind speeds (V_G = 25-50 m/s i.e. 55-110 m.p.h.) and for latitudes 30° and 60°. It is seen that the ratio is relatively insensitive to latitude or gradient wind velocity but highly sensitive to the roughness of the surface. Furthermore, the ratios of wind speed at the two reference heights agree quite well with the ratios suggested in the simple power law profiles of Figs. 4 and 5. The great advantage of the presentation of Eq. 13 and Fig. 9 is that due to the insensitivity to the latitude and wind speed the ratio $\frac{V_Z}{V_G}$ is determined almost uniquely by the roughness length Z_o. For middle latitudes and more usual extreme gradient wind speeds, equation (13) might be written

$$\frac{V_Z}{V_G} = 0.082 \, Z_o^{0.09} \ln \frac{Z}{Z_o} \qquad (14)$$

3.2 Advection

In the previous section it was tacitly assumed that the surface of the ground was uniform for a sufficient distance upwind for steady-state conditions to be established. This is obviously not always so and in many instances such as the outskirts of cities the wind profile goes through a transition. An important question is how far down wind of a change in roughness is it before the wind regime appropriate to the new surface is established. Jensen's results give a clear indication of the order of the variation but are insufficiently detailed to be precise.

The problem has however been treated by Taylor[32] who determined approximately the fetch distance downwind. His results are shown in Fig. 10. The parameter r denotes the roughness length ratio $\frac{Z_o^I}{Z_o^{II}}$. For a typical example of a wind blowing from flat open terrain ($Z_o^I \approx 10$ cm) to a suburban area ($Z_o^{II} \approx 100$ cm), $r = 0.1$. For the wind regime appropriate to a suburban area to establish itself at 100 ft requires a downwind distance of approximately 3/4 mile from the change in roughness. At a height of 300 ft the distance required is about 5 miles. For a wind blowing in the opposite direction (from suburban area to open country) $r = 10$, and the downwind distance required for a 100 ft height is about 2 miles and for 300 ft about 9 miles. Thus it takes almost twice as far for the wind to pick up speed over the smoother surface as it does to slow down over the rougher.

Meteorological stations are in many instances located at airports and the assertion is often made that the wind there is typical of flat open country. The results given above suggest that if the airport is situated on the outskirts of a city for this to be true there should be a clear fetch of some 2 or 3 miles. This is seldom the case even in one direction.

Fig. 10 is clearly of importance in estimating any reduction in mean wind velocity which might be made on account of the rougher nature of the terrain.

3.3 Orographic effects

The orographic effects on the wind (due to mountains and valleys) are extremely difficult to define with generality. The principal effects are the amplification which can arise at mountain tops to the funnelling of wind in valleys. This is generally extremely local and the wind profile and velocity amplification can in most cases only be satisfactorily established by site investigation (see overleaf).

4. CHARACTERISTICS OF GUSTINESS NEAR THE GROUND IN HIGH WINDS

4.1 General

An erroneous convenience commonly adopted in wind loading is that the wind is comparatively steady and hence gives rise to steady pressures. In fact, storm winds are usually extremely unsteady and turbulent: and rates of change of over three hundred miles per hour per second over ranges of 20 or 30 m.p.h. have been measured.

While there is still no completely satisfactory definition of turbulent flow one of its most recognizable features is its randomness. Because of this the description of a turbulent flow reduces to a description of its statistical properties. Typically, a turbulent flow at a point is characterized by a mean velocity and three fluctuating components of velocity in three mutually perpendicular directions. Two types of statistical measure are important. First the probability function defining the distribution of velocities for each component. Second the correlations within the turbulent field between the velocity components. Alternative to and interchangeable with the correlation functions are the spectra and cross-spectra of velocity which define the contributions made to the variances and covariances of velocity by fluctuations of different frequencies. To completely define the flow through these quantities is a formidable task particularly in the boundary layer where isotropy extends at most to the horizontal. From the practical point of view however, only some of these quantities seem important. These are,

1. The spectrum of horizontal wind speed
2. The spectrum of vertical velocity
3. The spatial correlations of the velocity components at specific frequencies.

These and the probability density are now discussed.

4.2 Probability density

The probability distribution of wind velocity has been examined[3,5,17,28,30] and shown to agree well with the familiar normal or Gaussian distribution given by

$$p(v) \cdot dv = \frac{1}{\sqrt{2\pi}\sigma} \cdot e^{\frac{-(v-\bar{v})^2}{2\sigma^2}} \cdot dv \qquad (15)$$

where σ^2 is the variance of the velocity (mean-square fluctuation). From a study of 1 second wind velocity measurements taken over a large number of separate five-minute periods on a 350 ft mast Huss and Portman[18] concluded: "The assumption of a normal frequency curve for representing wind-velocity distribution, while not accurate for any given case, would appear to be justifiable for a large number of cases."

4.3 Spectrum of horizontal gustiness in high winds

The spectrum of horizontal wind speed over an extended frequency range has already been referred to in Fig. 2. Attention is now confined to the high-frequency end of this spectrum pertaining to gusts. From a study of some 90 strong-wind spectra obtained at different heights, surfaces and parts of the world, the writer[10] suggested the following expression for the spectrum of horizontal gustiness.

$$\frac{n \cdot S(n)}{V_*^2} = \frac{n \cdot S(n)}{\kappa \bar{V}^2} = 4 \cdot \frac{x}{(1 + x^2)^{4/3}} \qquad (16)$$

where $S(n)$ is the spectrum of horizontal speed at frequency n and height z and $x = 4000 \frac{n}{\bar{V}_1}$ where $\frac{n}{\bar{V}_1}$ is in waves per foot.

This curve is illustrated in Fig. 11 together with the averaged results for a number of different localities and heights, some of which were given previously. Additional results are given for the Severn River Railway Bridge[11] and for a 150 ft tower near the centre of London, Ontario. The individual spectra for the latter are given in Fig. 12. The city here is situated in flat country and the tower stands 150 ft high amidst houses and larger buildings up to four stories high along wide avenues lined with trees up to fifty feet high (mainly deciduous). The relationships between drag coefficient and wind profile already outlined suggests that the appropriate value of the drag coefficient κ is 0.030-0.035 and the roughness length $Z_o = 1$ m. If so, this site is the roughest for which spectral measurements are available.

It appears that there is quite satisfactory consistency between the various curves for all types of surface and heights above ground (some variation can be attributed to variations in instrument response and method of analysis). Three particular features are noteworthy about the spectrum. First, the peak which occurs at a wavelength of approximately 2000 ft, second, the proportionality of the spectrum to $\left(\frac{n}{\bar{V}}\right)^{-2/3}$ for large values of $\frac{n}{\bar{V}}$ (a theoretical result) and third, the proportionality to the shear stress and friction velocity V_*^2. Since the mean-square fluctuation velocity is proportional to the area under the spectrum it follows from that the R.M.S. intensity of turbulence at height Z is

$$I_Z = \frac{\sigma}{\bar{V}_Z} = \frac{\left[\int_0^\infty S(n) \cdot dn\right]^{1/2}}{\bar{V}_Z}$$

$$= \sqrt{\frac{6 V_*^2}{\bar{V}_Z}} \qquad (17)$$

$$= 2.35 \sqrt{K} \frac{\bar{V}_1}{\bar{V}_Z}$$

4.4 Spectrum of vertical gustiness

The spectrum of vertical gustiness has been studied extensively by Panofsky and McCormick[25]. A suggested empirical formula is

$$\frac{n \cdot S_W(n)}{\bar{V}_*^2} = 6 \frac{f}{(1 + 4f)^2} \qquad (18)$$

where $f = \frac{nz}{\bar{V}}$, the ratio of the height to wavelength. This curve is illustrated in Fig. 13.

Again, a peak is apparent when $\frac{nz}{\bar{V}} = 0.25$, that is, when the wavelength $\frac{\bar{V}}{n}$ is four times the height. An important distinction between vertical and horizontal gustiness is that the former appears to be strongly dependent on the height z.

4.5 Vertical and lateral correlations

When considering the wind loading on an extended structure such as a long bridge, tall mast or skyscraper it is clearly important to have some measure of the spatial distribution of gusts. This can be measured by the correlation coefficient between two velocity measurements spatially separated by the interval Δx. The co-variance between the two velocity measurements is the average product $\overline{V_1 V_2}$ and the cross-correlation is $\frac{\overline{V_1 V_2}}{\sqrt{\overline{V_1^2} \cdot \overline{V_2^2}}}$. Just as the variance could be broken down frequency by frequency into a spectrum, so can the co-variance (or cross-correlation). Following this approach we can define a "narrow band" correlation function or (cross-correlation spectrum) as

$$R_{\Delta x}(n) = \frac{S_{12}(n)}{\sqrt{S_{11}(n) \cdot S_{22}(n)}} \tag{19}$$

Unlike the spectrum itself, the cross spectrum $S_{12}(n)$ can be a complex quantity having both in-phase and quadrature components. The existence of the quadrature component can be taken to indicate a preferred orientation of eddies and therefore only occurs when there is asymmetry present in the flow. For example there is no significant quadrature component in the crosswind horizontal cross-spectrum between like components of velocity: in the vertical direction however, where there is strong asymmetry, the quadrature component is significant and the maximum correlation between the horizontal wind speed at two different heights occurs not simultaneously, but when the signal from the lower station is delayed by a time roughly equal to $\frac{\Delta z}{\bar{V}}$ where $\bar{V}$ is mean velocity: or where the instruments are positioned on a line inclined at 45° to the downwind direction.

The square of the absolute value of the cross-correlation spectrum is termed the "coherence." For practical purposes it is probably quite adequate to neglect the quadrature correlation, as such, and take the cross correlation as equal to the square root of the coherence.

The general form of the cross-correlation spectrum can best be seen from measurements made in a wind tunnel of the crosswind correlation of wind speed some distance downstream of a plane jet[12]. These are shown in Fig. 14. The cross correlation is expressed as a function of the non-dimensional ratio $\frac{n.\Delta x}{\bar{V}}$. Within the limits of the experimental accuracy the cross correlation can be quite well expressed by

$$R\left(\frac{n.\Delta x}{\bar{V}}\right) = e^{-C \frac{n.\Delta x}{\bar{V}}} \tag{20}$$

where $\quad C \approx 8$

This type of curve appears to be characteristic also of correlations in atmospheric turbulence. Fig. 15 shows measurements of the vertical cross-correlation ($\sqrt{\text{coherence}}$) of horizontal wind speed on masts situated in (a) open grassland, and (b) wooded terrain. Exponential decay curves similar to the above are shown in which the appropriate values of the coefficient C are roughly 7.7 and 6.0 respectively - remarkably close to the value found for the wind tunnel.

The quantity $\frac{1}{C} \cdot \frac{\bar{V}}{n}$ in fact defines the "scale" of the correlation (the distance to the centre of gravity of the correlation diagram or the "effective gust width"). Some valuable measurements of the along-wind and cross-wind scales of the along-wind u and cross-wind velocity components for specific wavelengths have been published by Cramer[7], see Fig. 16. These suggest that the along-wind scales of both the along-wind and cross-wind velocity components are roughly 1/6 of the wavelength in both stable and unstable atmospheric conditions. In unstable conditions the transverse scales of the same two velocity components are again roughly the same but slightly smaller being 1/10 of the wavelength. In stable conditions the cross-wind scales are very much less than the along-wind being roughly 1/40 and 1/25 of the wavelength for the along-wind and cross-wind components respectively.

The indication this gives is that in unstable conditions the along-wind and cross-wind scales are about equal (and equal to about 1/6-1/8 of the wavelength) in stable conditions the eddies are very much elongated in the direction of the wind and the cross-wind scales are of the order of 1/3-1/5 of the along-wind scale which itself is equal to roughly 1/8 of the wavelength. Which of these two models is representative of high winds is not yet clearly established although the weight of evidence is that the elongated eddy is more representative. The strongest evidence for this is probably in the directional traces for strong wind which do not tend to exhibit the wildly meandering characteristics of highly unstable conditions. In the boundary layer of the wind tunnel, Grant[15] who measured all nine correlation functions found that the longitudinal scale was 7 or 8 times larger than the lateral. This of course is also mechanical turbulence, as in strong wind. Some measurements of wind velocity across a broad front were obtained in 1937 by Bailey and Vincent[1] at the site of the Severn River railway bridge. These data were analysed by the writer and have been discussed by Panofsky[24]. From the cross-correlation coefficients between the wind at neighbouring instruments across the span the indications were that the cross-wind scale was somewhat less than 1/3 of the along-wind scale.

In previous papers the writer has suggested that because of the uncertainty, the assumption leading to the higher wind loading should be made and the cross-wind scale taken equal to the along-wind. This may now seem unnecessarily conservative and a somewhat smaller cross-wind scale might be adopted.

To summarize it is suggested that the following may be taken as representative values of the scales of turbulence at specific wavelengths:

$$\left. \begin{array}{l} \text{Vertical scale of wind-speed} \quad \frac{1}{7} \cdot \frac{\bar{V}}{n} \\ \text{Lateral } \\ \text{Lateral " " vertical velocity} \end{array} \right\} \frac{1}{25} \cdot \frac{\bar{V}}{n} \qquad (21)$$

4.6 The structure of turbulence

Perhaps the simplest notion concerning the structure of turbulence near a plane but rough surface, is that of a series of "roller type" eddies. This form has been tentatively suggested by Webb[37] for the natural wind and by Townsend[35] for the outer layer in the wind tunnel. Two models along these lines can be envisaged - one in which the eddies at the surface cartwheel along the ground in the direction of the flow and rotate in the same sense and one in which consecutive eddies rotate in opposite senses as indicated by Webb. The former poses the difficulty however that the neighbouring boundaries of the eddies must be moving in opposite directions (one up and one down) with consequent high shear. While the latter could well be typical of convective conditions in which thermal updraft provides the energy for rotating neighbouring cells of air in opposite senses it is difficult to see how such an eddy system could be generated mechanically (as in high wind), since it implies surface motion first opposing the flow then with it. Further it is difficult to see why, even if the turbulence is generated convectively the eddies should prefer to be lined up in the wind direction any more than across wind. In fact the similarity of longitudinal and across-wind scales in unstable conditions (see Fig. 15) suggests they do not have any preferential orientation.

More recently Townsend[34] has quoted some wind-tunnel measurements which are not generally consistent with the "roller" hypotheses. Instead he suggests that the motion consists of two dimensional jets which originate in the immediate neighbourhood of the wall surface. Such jets with their surrounding induced flow are shown diagrammatically in Fig. 17. Within the jet, the streamlines are deflected down wind, by the action of the velocity profile. The slower longitudinal velocity in the jet compared to that outside is due to the transport upwards of the slower moving fluid near the surface. In three dimensions the flow within the jet and the return flow outside resembles in some senses two contra-rotating corkscrews with their axes lying parallel to the mean flow and their inside boundaries both moving upwards, constituting the jet flow.

There are several features of this model which are supported by observations in strong wind. First it accounts for the displacement of the maximum correlation of the longitudinal velocity component in the vertical direction to a position along a line inclined at roughly 45° to the mean flow and coinciding approximately with the mean of the postulated jets. (See discussion on scales.) It accounts for the much larger longitudinal scale compared to the lateral scales. Third it accounts for the influence of height on the vertical scale but its apparently much weaker influence on the horizontal scales. (This is implied by the dependence of the vertical spectrum on the parameter $z\frac{n}{V}$, z being the height, whereas the wind-speed spectrum depends on $L\frac{n}{V}$ where L is a horizontal scale found to be of the order of 4000 ft). Finally this model does not incur high shearing either at the eddy boundaries or at the ground surface - as does the roller model. This in a sense implies that the air flow is taking a path of least resistance or obeying the minimum energy principle.

Before firm theories can be put forward however it would seem desirable to have more information.

5. SYNTHESIS

The discussion of the wind structure so far can be summarized as follows. The wind is caused by the presence of large weather-map-scale disturbances which give rise to air movements having characteristic periods of the order of hours and velocities given by the gradient (or geostrophic) velocity. Nearer the ground the roughness of the surface sets up shear forces between the air and the ground which retard the flow throughout the planetary boundary layer. The mean-wind profile in this layer, which approaches the gradient-wind velocity at the top of the layer, can be expressed by power law or other profiles which only depend significantly on the characteristic roughness of the surface. The surface wind speed depends on the roughness and can vary greatly over short distances. In strong winds the boundary layer is fully turbulent. The shear stress at the surface induces mechanical turbulence the energy of which is proportional to the shear stress and depends significantly on the roughness of the surface and the height above the surface. The characteristic frequencies of the mechanical turbulence or gusts are altogether different to those of the weather-map fluctuations and a gap appears to exist in the spectrum of wind speed centered at a period of roughly one hour. This justifies the use of mean hourly wind speed, or somewhat shorter averaging periods as the characteristic velocity. The wind speed and vertical gust spectra can be expressed to an adequate accuracy knowing only the mean hourly velocity and the roughness of the surface.

All of this boils down to the simple fact that almost all the properties of the wind that might be needed in structural design can be estimated reasonably accurately provided that the mean-wind field and the ground roughness are known.

5.1 The mean-wind field

The properties of the mean-wind field of major interest in structural engineering relate mainly to the extremes and their recurrence intervals. Another property, the overall statistical distribution defining the proportion of the total time the wind lies between certain values would also be needed in problems dealing with fatigue (also wind power), but it may be several years before such information is generally applied. (Some illustrations of these distributions have been given by Tagg[31].

There would seem to be two approaches in defining the wind field, one is to define the surface wind field itself, the other, to define the gradient wind and from it estimate the surface wind from the roughness and the wind profile. Whichever method is used - and the relative merits will be discussed below - it seems that the only suitable data on which the wind field can be based are surface measurements from meteorological stations. Another possibility - the use of isobaric maps - has been investigated by the writer, but does not appear promising. These are only available two or three times a day, and so do not constitute a continuous record. They are inevitably to some extent subjective - depending on the meteorologist who draws the map - and furthermore they are least accurate when the winds are strongest and the isobars most closely packed.

If the wind field is to be based on surface measurements then it may seem most logical to define it also in terms of a surface velocity at some arbitrary reference height rather than the gradient velocity. A difficulty that this approach presents is that, in general, changes in surface roughness take place on a far smaller scale than the grid spacing of meteorological stations (as Jensen's results well indicate). Interpolations are still needed therefore if adequate estimates are to be provided at all localities. A further difficulty lies in the reliability of the individual surface estimates themselves. The periods for which meteorological records are available varies widely; the heights at which the instruments are established vary; frequently they have been moved up and down and from place to place to make room for new buildings and in so doing the exposure has been changed; urban sprawl may have appreciably changed the effective roughness of the surroundings; at coasts the instruments are situated in regions of wind transition and the records are not typical of wind conditions a short distance inland, unserviceability may have broken the continuity of the records. For all these reasons some basis of comparability and cross checking of records seems essential. The most obvious basis of such a comparison is the estimate of the gradient wind made from surface records.

Unlike surface winds, the scale of variations of gradient wind are very much larger than the normal grid of meteorological stations and estimates from several stations should normally overlap. This, in fact, is an argument in favour of using the gradient wind to specify the mean-wind field - at least as an intermediate step. Whether or not it is finally stated in this form for use by structural engineers is another matter. Possibly the gradient wind is too nebulous a concept for down-to-earth engineers! On the other hand the fact that specification in terms of the gradient wind would require the engineer's participation in evaluating the exposure of the structure, and hence the wind loading itself may be of direct benefit in eventually arriving at realistic wind loading.

5.2 Extreme winds

The recurrence of extreme wind speeds can be estimated using extreme value statistics whereby the distribution of largest values in samples of given size can be estimated. It has been indicated above that the wind speed averaged over 1 hour (or thereabouts) is a key quantity in estimating wind loads. If so, what are needed are estimates of the probability of given values of hourly velocity occurring during periods commensurate with the lifetime of the structure. Another way of expressing this is to determine the recurrence interval for different values of hourly wind velocity. A useful time interval is the year. The application of extreme value analysis to wind loading has been discussed elsewhere by Court[8], Shellard[27], Gumbel[16], Thom[33], Jenkinson[19], Boyd and Kendall[4] and the writer[9]. Boyd and Kendall compared three different extreme value methods as applied to wind speed (the Fisher-Tippet 1 curve as used by Gumbel, the Fisher-Tippet 2 curve as used by Thom and A. F. Jenkinson's combined approach). The three methods differ mainly with regard the assumptions on the lower bound and sample size. They are compared with the observed distribution of extreme winds over 32 year period at Victoria, B.C. in Fig. 19. Boyd and Kendall conclude that although none of the methods is ideal for small samples, there is little reason for preferring the methods used by Thom or Jenkinson to the more straightforward method developed by Gumbel. Court analyzed the records of twenty-five weather stations in the United States having 37 years of satisfactory records by the Gumbel method and stated "all of the wind data seems to follow the theory."

The distribution function is of the form

$$F(x) = e^{-e^{-y}} \qquad (22)$$

where y is the reduced variate given by

$$y = a(x - U) \qquad (23)$$

a being the dispersion factor and U the mode. For return periods (r) greater than about 10 years, the corresponding value of x is

$$x \approx U + \frac{1}{a} \ln r \qquad (24)$$

The significant factors in this expression are a and U; knowing these, the wind velocity having any desired return period may be estimated. The territorial variation of U and $\frac{1}{a}$ can in a sense be taken as defining the extreme-mean-wind-velocity field. The U and $\frac{1}{a}$ values defining the extreme-gradient-velocity field can be estimated from surface values using these relationships such as (8) and (14) provided the ratio of surface to gradient wind speed can be assessed. In an earlier paper the writer[9] attempted this for the British Isles using values of U and $\frac{1}{a}$ estimated by Shellard[27] from surface records and estimating the ratio of surface to gradient wind speeds using estimates of ground roughness from descriptions of the stations and their surroundings. The results were reasonably encouraging but it was felt that much better could have been done if site inspection and field tests could have been made to determine the roughness more directly. A modified form of the original map is shown in Fig. 19. The surface wind is then determined from the wind profile most appropriate to the terrain and corresponding to the found value for the gradient wind.

The method used in allowing for the variations in surface roughness can be assessed from a comparison of extreme wind speeds at city and airport offices for various cities in the United States shown in Table 1. The data was kindly made available to the writer by Mr. H. C. S. Thom of the U.S. Weather Bureau. The wind speeds refer to the 2 per cent quantile (once in fifty years). As can be seen, in spite of the much greater anemometer height Z_A in the city the wind speed is substantially less (the average being about 70% of the airport speed). The actual ratios are seen to give not unreasonable agreement with the predictions made according to the earlier paper[9].

5.3 Site investigation

It is normal engineering practice for the design of important structures to be preceded by investigations of the various geophysical properties of the site, in particular the geology, the soil and ground water. The value of having detailed information on these pays for the cost of such investigation many times over. Site investigation of the wind conditions could prove to be no less useful and could be carried out at very little expense. The main purpose of such an investigation would be to determine the wind profile - and hence the ratio of surface to gradient winds - the roughness and the gustiness. They would be particularly valuable at hill-top sites where the more usual aerodynamic profiles may be considerably distorted. Such investigations have in fact been conducted in connection with wind-power sites.

The equipment needed would amount to no more than a portable mast with anemometers and field recorders, and pilot balloon equipment or smoke trail rockets would be adequate for most purposes. Records of strong wind of 15-20 miles/hour on two or three occasions would probably provide most of the information that a more protracted study would yield.

TABLE 1

COMPARISON OF EXTREME WIND SPEEDS
AT CITY AND AIRPORT STATIONS IN THE UNITED STATES

Station	City Office			Airport			City Office		Airport		Ratio: City/Airport	
	Z_A (feet)[1]	2% quant. @ 30 ft m.p.h.[1]	2% quant. @ Z_A m.p.h.[2]	Z_A (feet)[1]	2% quant. @ 30 ft m.p.h.[1]	2% quant. @ Z_A m.p.h.[2]	Roughness C'tgy[3]	K_A (C.O.)[4]	Roughness C'tgy[3]	K_A (A.P.)[4]	Actual	Predicted[5]
Boston	188	55	72	63	93	103	8	2.3	3	1.5	0.70	0.65
New Haven	155	47	60	42	70	74	7-8	1.85	4	1.81	0.81	0.98
Chicago		46	57	38	68	70	(Roughness difficult to estimate)					
S. S. Marie	52	54	63	33	84	85	7	3.0	4-5	2.1	0.74	0.70
Kansas City	181	48	63	76	83	95	7-8	2.3	3	1.5	0.66	0.65
Omaha	121	53	65	68	81	91	7-8	2.6	4	1.70	0.71	0.65
Knoxville	111	47	57	71	79	89	7-8	2.6	5	1.90	0.64	0.73
Nashville	191	55	73	42	82	86	7-8	2.3	4-5	1.95	0.85	0.85
Spokane	110	42	51	29	78	78	7-8	2.6	3	1.75	0.65	0.67

NOTES

1. From data provided by U. S. Weather Bureau.
2. Readjusted using 1/7th power law (see Ref. 1).
3. Estimated roughness category of anemometer environment (2-3 mile radius) in direction of prevailing wind from U.S. Weather Bureau descriptions using Table 2 of Ref. 2.
4. Ratio gradient to surface wind velocity (anemometer height) from Fig. 4, Ref. 2.
5. Ratio K_A (A.P.) to K_A (C.O.).

REFERENCES

1. Thom, H.C.S. "Distributions of Extreme Winds in United States", Paper 2433 *Proc. Amer. Soc. civil Engrs*, May 1960.
2. Davenport, A.G. "Rationale for determining design wind velocities", Paper 2475 *Proc. Amer. Soc. civil Engrs*, May 1960.

6. CONCLUSIONS

This paper has attempted to describe as far as possible the macro- and micrometeorological structure of strong winds in the earth's boundary layer insofar as it affects the wind loads on structures. It seems appropriate to conclude by referring to some of the practical implications.

This study has shown that extreme mean wind speeds at building height (say 30 ft) can vary widely. As well as the large-scale global variations in wind climate, much more local variations are also important. It has been shown that the roughness of an urban area itself may slow the wind down to a third of the speed in open country at the outskirts. The corresponding decrease in mean wind pressure (proportional to the velocity squared) is one ninth. In addition, due to different return periods for strong winds, it may be logical to design a monumental structure for a mean wind velocity up to 1½ times greater than that for a shorter life structure, corresponding to a mean wind pressure up to twice as large. These two influences alone (roughness and lifetime) call for a possible factor of 18 in the variation of mean wind loads. To this must be added (or subtracted) the effect of gusts. The wind in a city is much gustier than in open country (by a factor of three or four). Whether this is significant or not depends on the susceptibility of the structure to gusts, that is, whether the natural frequency of the structure coincides with the frequency of high energy fluctuations in the wind spectrum, whether the damping is high or low, and whether the scale of the gusts at critical frequencies is large or small compared to the structure. These considerations taken together suggest a range of values for a logically consistent and realistic set of wind pressure standards which at first sight may seem staggering and is in contrast to the blanket wind loading currently used.

It is fair to ask whether these variations are economically significant or only marginal. The answer is probably that a careful tailoring of wind loading would in many operations such as power transmission, long span bridging and tall skyscrapers and towers provide very substantial savings. Very often in such structures the benefits from small decreases in wind loads snowball rapidly. Smaller structural members are needed, often implying less exposed area to the wind and again smaller wind loads, implying lighter structures and less expensive foundations etc. etc.

The economic value of site investigations in achieving some of these possible savings still has to be fully appreciated. Moreover the savings (or additional utility) that a more careful statement of wind loads would yield, would pay for the necessary exploration and research countless times over.

What research seems necessary? Fundamental to the entire subject of wind loading is of course the careful mapping of the large-scale territorial wind field. Work on the estimation of extreme wind speed parameters at individual weather stations is a valuable step: much needs to be done however in the evaluation of the quality of the measurements themselves. A method for integrating and correlating the measurements into a wind field is suggested by the map given of extreme-gradient-wind parameters. This could be improved probably in a variety of ways, but principally by simple soundings of the wind profile at weather stations (for example by instrument tower, balloon or smoke rocket).

As suggested, the most useful parameter on which to base the wind field is the mean velocity taken over a period between five minutes and one hour (the difference between the two means is not likely to be great). Other measurements such as maximum gust speed are not highly meaningful quantities in relation to wind loads.

More detailed research still needs to be done in determining the wind structure over an urban area - its profile and spectrum. Some general work on spectra and scales of turbulence is in progress at a number of places and this may give rise to modifications to the values suggested here - but the consistent trend in the work so far suggests these may not be large.

REFERENCES

1. BAILEY, A. and VINCENT, N. D. G. "Wind pressure experiments at the Severn Bridge". *J. Inst. civ. Engineers*, 1939. 11, 363.
2. BAUMGARTNER, A. "Untersuchungen ueber den Waerme und Wasserhaushalt eines jungen Waldes". *Ber. dtsch. Wetterdienstes, U.S. Zone*, 1956. 5, (28).
3. BEST, A. C. "Transfer of heat and momentum in the lower layers of the atmosphere". *Geophys. Mems.* 1935. (65), 40.
4. BOYD, D. W. and KENDALL, G. R. "Statistics of extreme values". *Meteorological Division, Department of Transport of Canada*, 1956. CIR 2825, Tec-238.
5. CRAMER, H. E. "Preliminary results of a program for measuring the structure of turbulent flow near the ground". *Geophys. Res. Paper*, 1952. (19), AFCRC.
6. CRAMER, H. E. "Use of power spectra and scales of turbulence in estimating wind loads". *Unpublished (Paper presented at the 2nd National Conference on Applied Meteorology, 1958. Ann Arbor, Michigan.)*
7. CRAMER, H. E. "Measurements of turbulence structure near the ground within the frequency range from 0.5 to 0.01 cycles sec^{-1}". *"Advances in Geophysics", Academic Press*, New York, 1959, (*Proc. of Symp. on "Atmospheric Diffusion and Air Pollution"*, Oxford, 1958).
8. COURT, A. "Wind extremes as a design factor", *Frank. Inst.*, 1953. 256, (1).

9. DAVENPORT, A. G. "A rationale for the determination of design wind velocities". *Proc. Amer. Soc. civ. Engrs., J. struct. Div.*, 1960. **86**, 39.
10. DAVENPORT, A. G. "The spectrum of horizontal gustiness near the ground in high winds". *Quart. J. roy. meteor. Soc.*, 1961. **87**, 194.
11. DAVENPORT, A. G. "The buffetting of a suspension bridge by storm winds". *Proc. Amer. Soc. civ. Engrs. J. struct. Div.*, 1962. **88**, 233.
12. DAVENPORT, A. G. "Wind tunnel measurements of the fluctuating forces on rectangular discs and panels of an infinite flat plate due to turbulence". *Unpublished. (Report submitted to National Aeronautical Establishment, N.R.C., Ottawa, 1962. To be published with R. L. Wardlaw by N.R.C. shortly).*
13. DAVENPORT, A. G. "The response of slender line-like structures to a gusty wind". *Proc. Inst. civ. Engrs.*, 1962. **23**, 389.
14. DELAND, R. J. and PANOFSKY, H. A. "Structure of turbulence at O'Neill, Nebraska, and its relation to the structure at Brookhaven". *Penn State Univ., Div. Met. Phys.*, Sci. Rep. No. 2, 1957. U.S.A.F. Project 19 (604) 1027, AFCRC-TN-57-262.
15. GRANT, H. L. *J. fluid Mechs.*, 1958. **4**, (2), 149.
16. GUMBEL, E. J. "Statistical theory of extreme values and some practical applications". *Appl. maths. Ser. nat. Bur. Stand*, 1954. 33.
17. HESSELBERG, T. H. and BJORKDAL, E. "Uber das Verteilungsgesetz der Wind ruhe". *Beit. Phys. frei. Atmos.*, 1929. **15**, 121.
18. HUSS, P. O. and PORTMAN, D. J. "Study of natural wind and computation of Austausch turbulence constant". *Daniel Guggenheim Airship Inst.* Rep. No. 156, 1948. Akron Univ.
19. JENKINSON, A. F. "The frequency distribution of the annual max. (or min.) values of Meteorological elements". *Quart. J. roy. meteor. Soc.*, 1955. **81**, 158.
20. JENSEN, M. "Shelter effect investigations into the aerodynamics of shelter and its effects on climate and crops". *Copenhagen, Danish Tech. Press*, 1954.
21. LETTAU, H. H. "Wind profile surface stress and geostrophic drag coefficients in the atmospheric surface layer". *"Advances in Geophysics", Academic Press, New York*, 1959. (Proc. of Symp. on *"Atmospheric Diffusion and Air Pollution"*. Oxford, 1958).
22. JOHNSON, O. "The relation between wind velocity at 2M, 40 ft. and 1000 ft. and its dependence on atmospheric stability". *Ralston Alberta Suffield Exp. Sta., Suffield Tech.* 1958. Paper No. 116.
23. NUNNER, W. "Warmenbergang und druckabfall in rauhen rohren". *VDI-Forschungsheft* 455, 23. 1956. Vereins Deutscher Ingenieure.
24. PANOFSKY, H. A. "Scale analysis of atmos. turbulence at 2 m.". *Quart. J. roy. meteor. Soc.*, 1962. **88**, 57.

25. PANOFSKY, H. A. and McCORMICK, R. A. "The spectrum of vertical velocity near the surface." *Quart. J. roy. meteor. Soc.*, 1960. 86.
26. ROSSBY, C. G. and MONTGOMERY, R. B. "The layer of frictional influence in wind and ocean currents". *"Papers in Physical Oceanography and Meteorology"*, *M.I.T. and Woods Hole Oceanographic Inst.*, 1935. III, (3).
27. SHELLARD, H. C. "Extreme wind speeds over Great Britain and North Ireland". *Meteor. Mag.*, 1958. 87, 257.
28. SHIOTANI, M. "Turbulence in the lowest layers of the atmosphere". *Sc. Rep. Tohoku Univ. Ser. 5, Geophysics* 2, 1950. (3), 167.
29. SINGER, I. A. "A study of the wind profile in the lowest 400 feet of the atmosphere". *Prog. Report No. 5 and No. 9, 1960-61. Brookhaven National Laboratory, Long Island, U.S.A.*
30. SUTTON, O. G. "Micrometeorology". *1st ed., 1953. McGraw-Hill Book Co. Inc.*
31. TAGG, J. R. "Wind data related to the generation of electricity by wind power". *E.R.A. Report*, 1956. C/T 115.
32. TAYLOR, R. J. "Small-scale advection and the neutral wind profile". *J. fluid Mechs.*, 1962. 13, 529.
33. THOM, H. C. S. "Distributions of extreme winds in the United States". *Proc. Amer. Soc. civ. Eng. J. struct. Div.*, 1960. 85.
34. TOWNSEND, A. A. *Proc. I.U.T.A.M. symposium on boundary layer research* 1957. Springer.
35. TOWNSEND, A. A. "Structure of turbulent shear flow". *Cambridge Univ. Press*. 1956.
36. VAN DER HOVEN, I. "Power spectrum of horizontal wind speed in the frequency range from 0.0007 to 900 cycles per hour". *J. Meteor.*, 1957. 14, 160.
37. WEBB, E. K. "Autocorrelations and energy spectra of atmospheric turbulence". *C.S.I.R.O. Aust. Div., Met. Phys.* Tech. Paper No. 5. 1955.

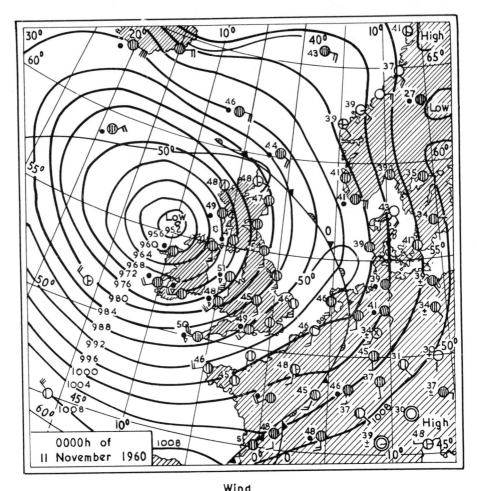

Fig.1. Weather map showing an intense depression with strong winds.

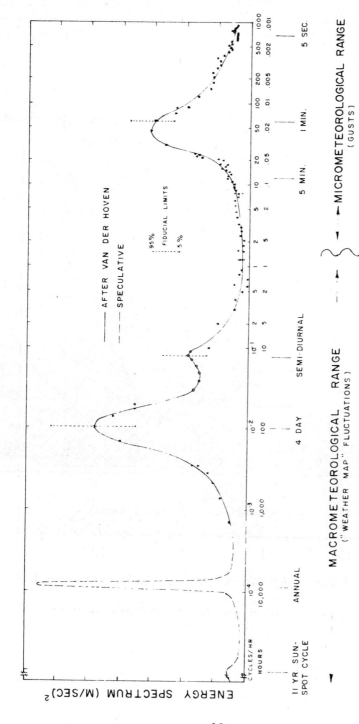

Fig.2. Spectrum of horizontal wind speed near the ground for an extensive frequency range (from measurements at 100 metre height by Van der Hoven at Brookhaven, N.Y., U.S.A.)

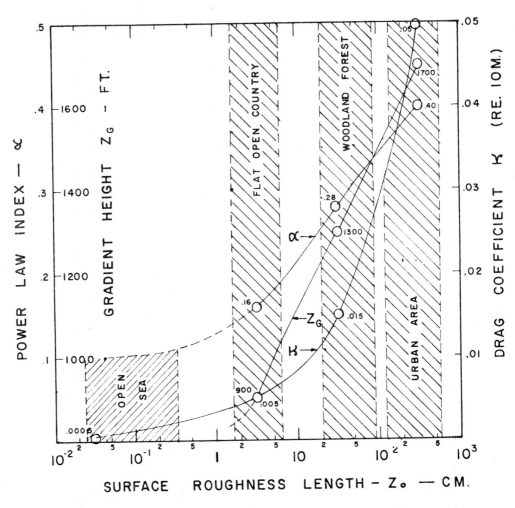

Fig.3. Parameters of wind profile for different surfaces.

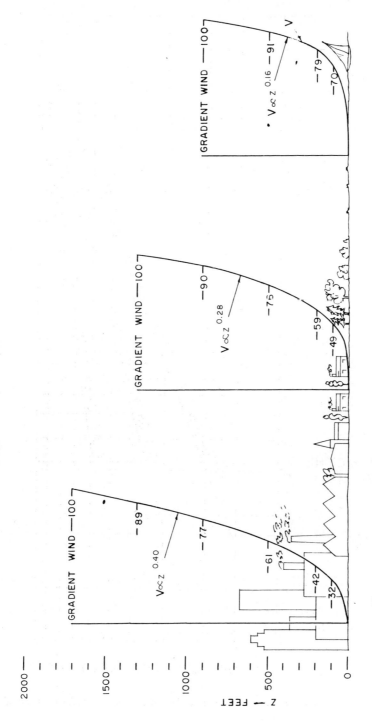

Fig. 4. Profiles of mean wind velocity over level terrains of differing roughness.

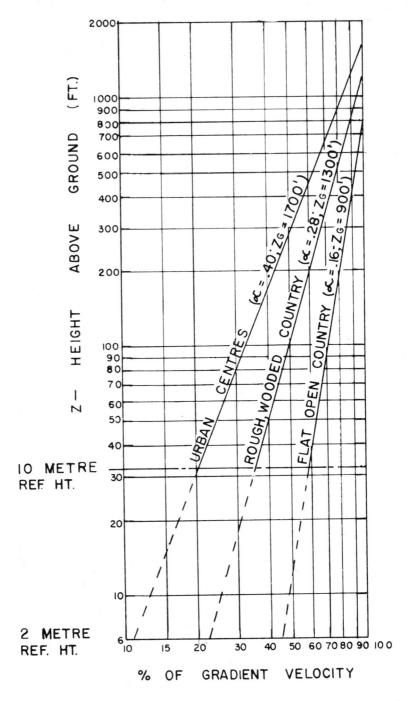

Fig.5. Power law wind velocity profiles for surfaces of different roughness.

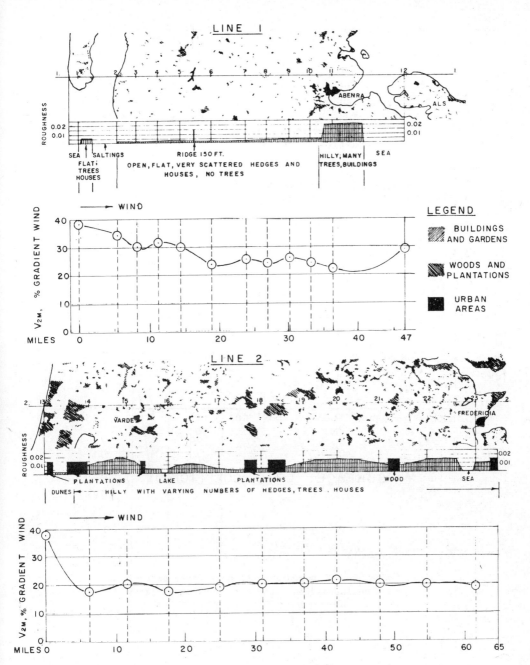

Fig.6. Jensen's measurements of the large scale variations in surface wind velocities with ground roughness (as indicated by the roughness of great order).

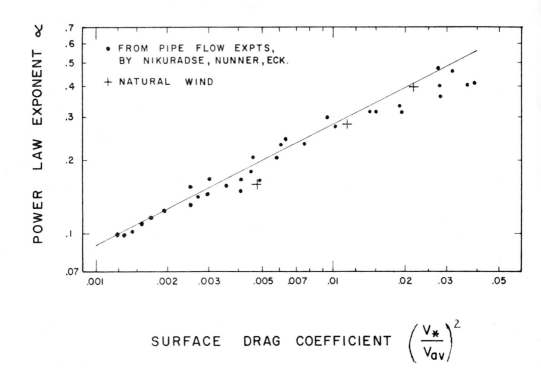

Fig.7. Comparison of turbulent boundary layer parameters in pipeflow and for wind over natural surfaces.

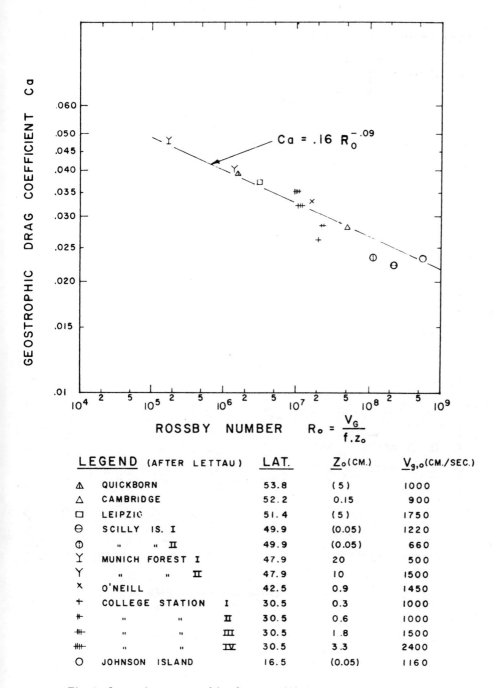

Fig.8. Lettau's geostrophic drag coefficient as a function of Rossby number.

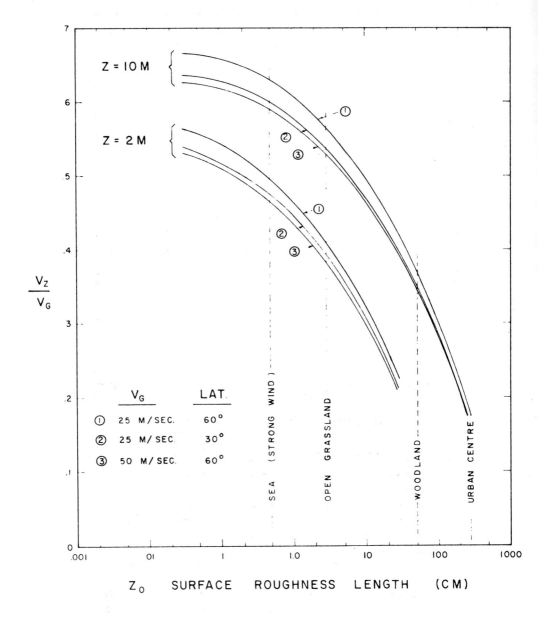

Fig.9. Ratio of surface wind speed to gradient wind speed as function of roughness length.

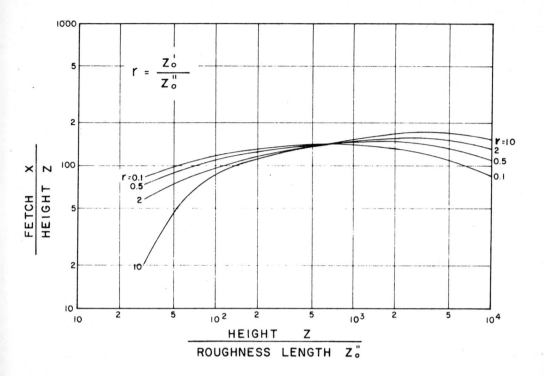

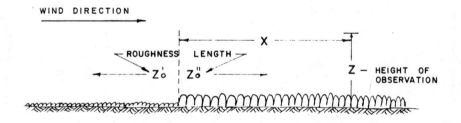

X = FETCH DISTANCE DOWNWIND OF CHANGE IN ROUGHNESS NECESSARY FOR ESTABLISHMENT OF WIND PROFILE APPROPRIATE TO NEW ROUGHNESS UP TO HEIGHT OF OBSERVATION.

Fig.10. Estimates of fetch distance necessary to establish new wind profile after change in roughness (after Taylor).

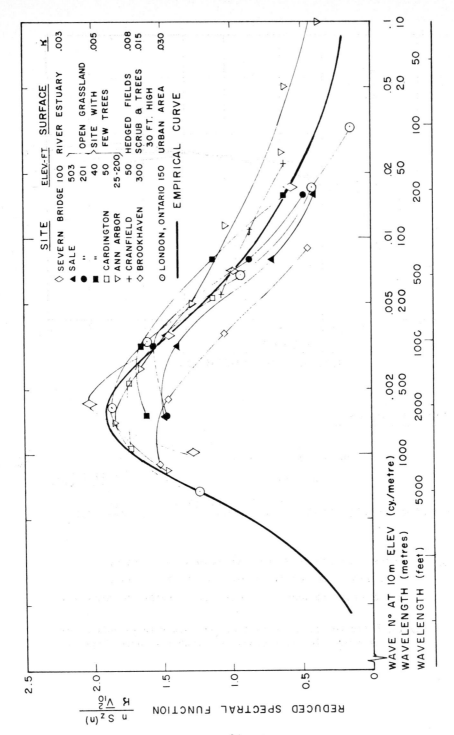

Fig.11. Spectrum of horizontal gustiness in high winds.

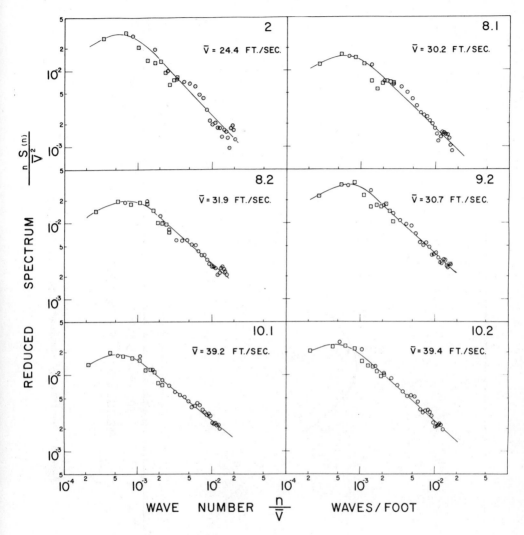

Fig.12. Spectra of horizontal wind speed at 150 ft. in strong wind over urban area (London, Ontario).

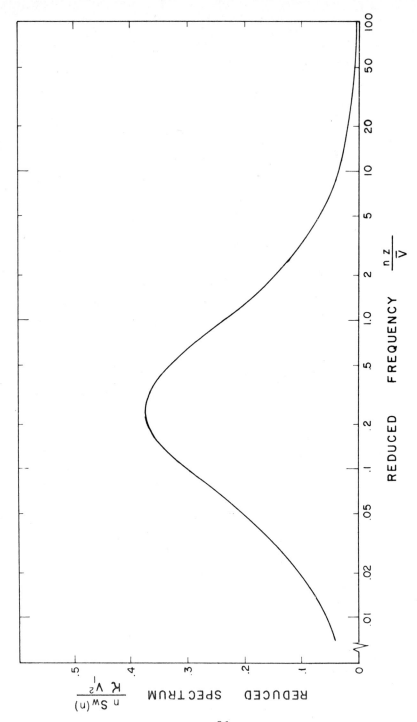

Fig.13. Spectrum of vertical gustiness (after Panofsky).

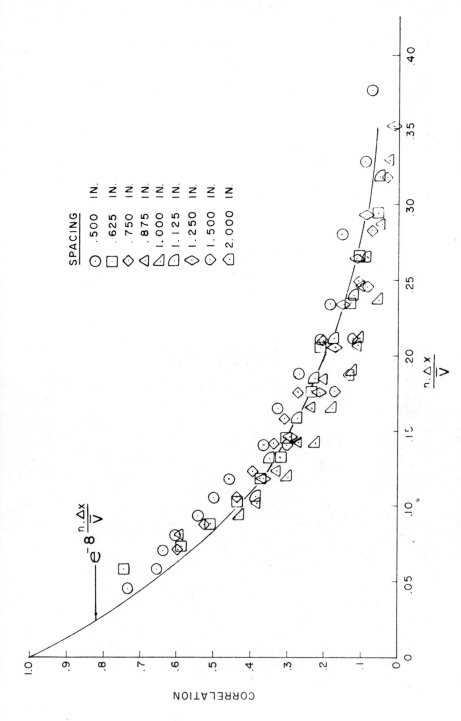

Fig.14. Lateral correlation of longitudinal velocity in strong turbulence in wind tunnel.

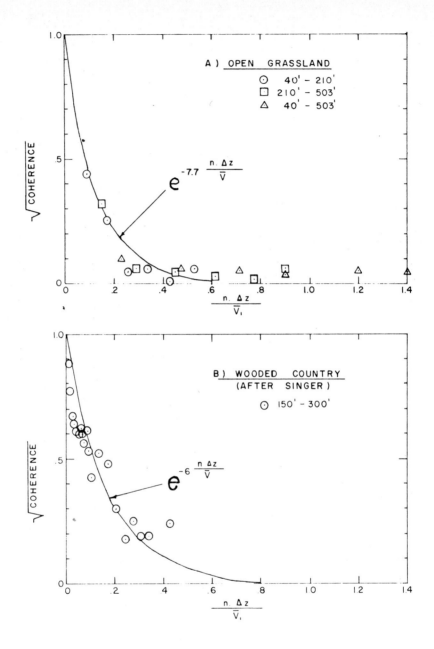

Fig.15. Absolute value of correlation (coherence) of wind speed in vertical direction as function of separation to wavelength ratio.

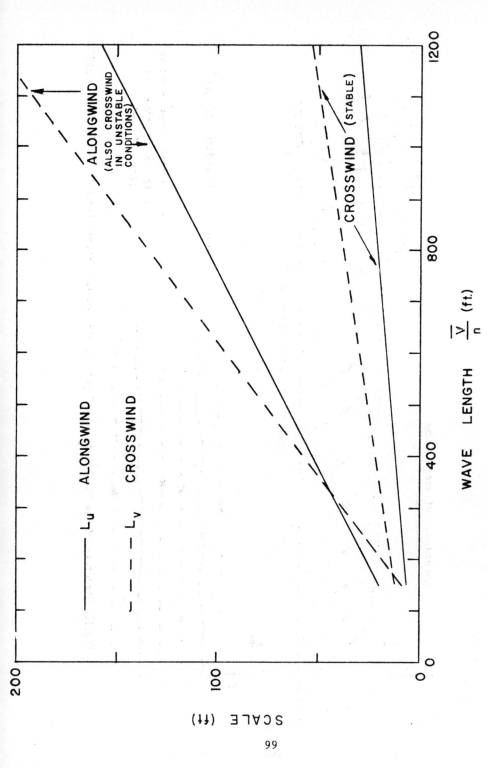

Fig.16. Along-wind and cross-wind scales of turbulence for the U- and V- components of wind velocity as functions of inverse wave number. (After Cramer).

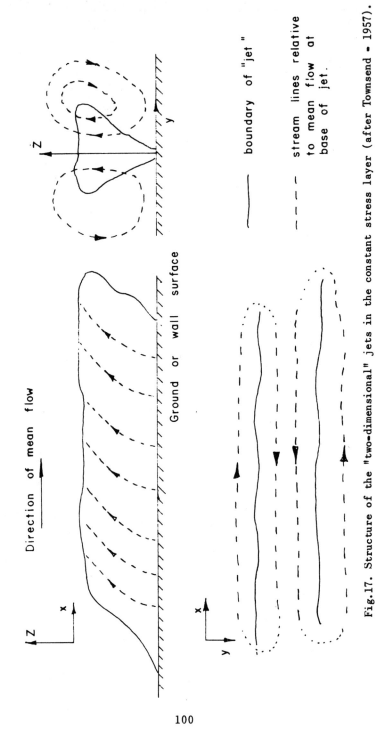

Fig.17. Structure of the "two-dimensional" jets in the constant stress layer (after Townsend - 1957).

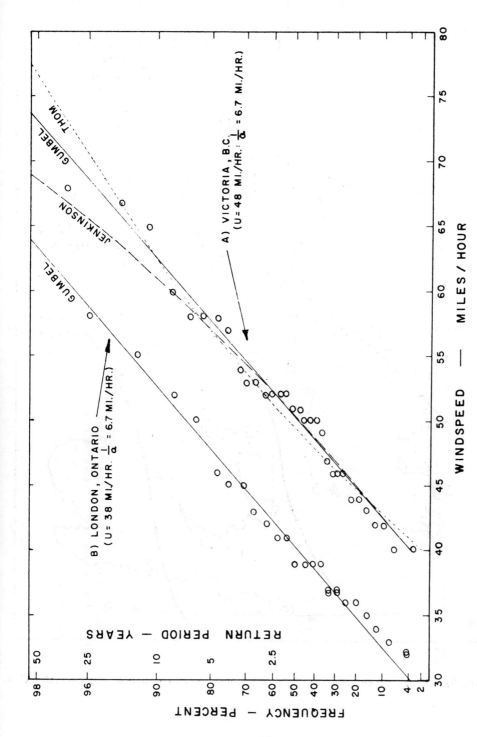

Fig.18. Comparison of annual maximum hourly wind speeds at Victoria B.C. and London, Ontario with theoretical distributions of extreme values (after Boyd and Kendall).

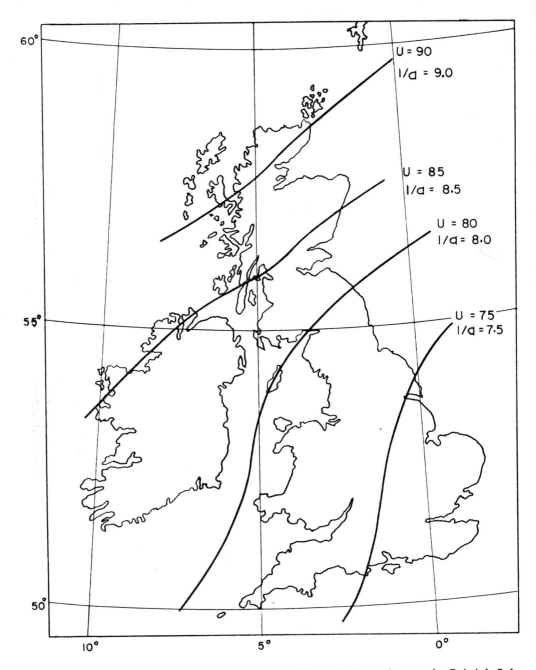

Fig.19. Parameters of extreme mean hourly gradient wind speed over the British Isles.

Units: Miles per hour

DISCUSSION ON PAPERS 1 AND 2

MR. GOLDING explained that the Electrical Research Station is concerned with wind speeds and with wind pressure. From 1948 it had studied wind speeds on especially windy sites in Great Britain and Ireland with the object of finding sites for wind-driven electrical generators. More recently it had been concerned with research on wind effects on very tall structures. While reliance will have to continue to be placed on Meteorological Office data, care must be taken in the treatment of the data. It is highly dangerous to draw the isovent maps which Mr. Shellard gives in *figs.3* and *4*. They would, for instance, indicate that Shetland and Orkney are no windier than Pembrokeshire whereas the E.R.A. measurements showed that Shetland and Orkney were much windier. It was possible to select sites with wind speeds of 140 m.p.h. whereas the isovent map predicted only 102 m.p.h.

PROFESSOR PAGE said that he would be disturbed if Professor Davenport's treatment of wind in cities was adopted. His experience during recent gales in the Sheffield area convinced him that the short period gust of 1 or 2 seconds was the most important one from the aspect of design, and that one minute duration was far too long. He thought that Professor Davenport's gradient for cities would underestimate the gust speeds in city centres, and that more attention should be paid to the effects of local topography such as the steepness of sloping ground. Another important aspect of wind loading is the formation of lee waves. It was the lee waves over the Pennines which caused considerable damage in Sheffield in the gale of 1962. Professor Page then showed a series of slides which showed the pattern of damage on the lee slopes of the Pennines. Examination of aerial photographs of tree damage after the gale showed that there was a great deal of damage on the lee slopes. Damage was often concentrated where a valley ran down in the same direction as the wind. An anemometer record showed a very high degree of gustiness.

PROFESSOR HENDRY supported Professor Page's suggestion that it was the 1 or 2 second gust which was significant in relation to cladding and building features and fixtures. With regard to the main load bearing structure, however, it seemed that a longer period could be justified. He referred to a paper by Professor Horne which dealt with the correlation of wind with the plastic behaviour of structures. He thought that this approach should be followed up, especially if more accurate knowledge of wind loading is linked with a closer look at the factors of safety and the load factors. Both authors had pointed to the desirability of further studies of wind over cities. Dr. Wilson at the University of Liverpool was undertaking one such study, with the aid of a captive balloon, up to heights of about 1,000 feet.

MR. SACHS mentioned some wind-speed measurements taken over short periods at sites proposed for the erection of large aerials. The ratio of the peak monthly gust at two stations varied between 1.18 and 1.62. This range was felt to be too large for the mean to be taken and so statistical theory was used on the hourly records for each month. The procedure narrowed the range from 1.21 to 1.38, and the value 1.38 proved to be fairly accurate for predicting the maximum gust at the second station from that of the first. Similar methods used to predict maximum mean hourly speeds proved unsatisfactory.

MR. RIMMER referred to some wind measurements Professor Davenport had made on a 900 foot tower. He did not understand how the fact that the speed was increasing at 66 feet and decreasing at 900 feet fitted in with a velocity profile as given in *fig.4*.

MR. ENTWISTLE referred to the discussion by Professor Hendry concerning structural failures and pointed out that South Yorkshire was littered with fallen floodlight pylons after the recent gale. These incidents raised the question of the significance of gusts in relation to the natural frequency of the structure. A major structure, carrying large super-loads, stands up to wind much better than lighter structures which have less super-loads.

MR. RICHARDSON thought that the curves shown by Professor Davenport relating to the spectral distribution of energy in the wind, and to the lateral and vertical correlations of the various components of the wind were very significant to the design of tall, slender structures and such information will eventually lead to a sound basis for design. However, even when sufficient of such information is available, information will still be required on the damping characteristics of the structure, for damping will become an important factor in design.

MR. SHELLARD (in reply) agreed with Mr. Golding on the dangers inherent in using the maps given in his paper. He had stressed this himself and had pointed out that these maps did not cover topographical effects.

PROFESSOR DAVENPORT (in reply) said that most of the points raised in connection with his Paper related to gust effects and these he thought might possibly be covered in his Paper 9 to be presented later. In reply to Mr. Rimmer the velocity profiles shown in his Paper referred only to means taken over 5 minutes or more. Mean speeds over shorter periods would not be stable and the condition of wind increasing at the bottom of the mast while decreasing at the top is not surprising since the correlation over the height for short gusts is very poor. He was not in disagreement with Professor Pages' remarks on wind loading in cities. He himself had made the point that gust intensity in cities was much greater than elsewhere.

WRITTEN DISCUSSION

DR. WYATT (Imperial College) wrote that he had found the first paper, by Mr. Shellard, to be very interesting on conjunction with Professor Davenport's paper 2. The mathematical description of wind structure proposed by Davenport can be analysed by methods he has described previously to compare with Shellard's Table IIa:

Ratio of max. speed averaged over time t to hourly mean			
t	20 sec	5 sec	½ sec
from table IIa	1.36	1.47	1.59
from postulated spectral distribution open country $K = .005$ $= 100$ f/s	1.26	1.46	1.66

The postulated spectral distribution is a generalization based on many observations, including probably some of those used by Durst. It also implies that the addition to be made to the mean wind speed to allow for gusts is independent of the height above ground (eqn. 17 of paper 2) and that the addition is only weakly dependent on the terrain: for example, taking the three categories of roughness considered by Davenport (*figs.3 and 4* of paper 2), for a gradient speed of 170 ft/sec corresponding to an open-country hourly mean speed at standard height of 100 ft/sec, the additions to allow for a 5 second gust are 46, 49 and 57 ft/sec respective to power law indices of 0.16, 0.28 and 0.40. As mentioned by Shellard the last category is extreme, at any rate for the building proportions of a British city.

Instead of different families of gust factors for each terrain together with different power law indices for gust speeds, the writer therefore proposes that a "flat rate" addition should be specified for each gust duration as a factor of the "open country" speed, irrespective of height or terrain. From the above example, to obtain the 5-second gust speed the addition would be 50% of the open country $\bar{V}_{10}$ or 30% of V_G. The 0.085 index for gust speeds in open country gives only small variation of gust addition with height, but the appropriate indices for other terrain have not attracted much attention: the proposed method avoids use of these indices. The suggested gust factors at the end of section V of paper 1 also correspond quite closely to addition of the same proportion of the open country speed, assuming the mean speed in the city to be about 60% of the open country value.

The results quoted by Shellard for mountain top conditions are most important as information of this type has been lacking hitherto. The relatively low gustiness is most striking.

	Observed max.^{m} speed		Gust predicted from mean	
	Mean	Gust	Same addition as at low level	Same gust factor as low level
Lowther Hill	47.6	69.2	65.3	79.7
Drum	50.7	68.4	63.0	74.0

The average of the two predictions shown fits the observations as well as inventing a new gust factor.

Another subject not previously described to engineers is the consideration of fetch in relation to effective exposure. Davenport gives important information here, but the writer would like to know if the dependence on fetch was asymtotic, and if so whether substantial readjustment to a new profile was likely in much shorter distances than those quoted in section 3.2 of paper 2.

PROFESSOR DAVENPORT. Structural engineers in Great Britain are fortunate in having perhaps the most well-developed and closely spaced network of first-class meteorological stations in the world. They are doubly fortunate in having had the data which has been accumulated over the years from this network, analyzed and compiled for them by Mr. Shellard. The benefits to be derived from this analysis are considerable when measured in terms of improved safety and economy in structures.

At the same time, the writer is disappointed by Shellard's negative reaction to the gradient wind approach, suggested previously by the writer as a way to define the territorial distribution of extreme winds. The principal advantage of this approach (which Shellard does not explain fully) is that it allows the heterogeneous collection of data from existing meteorological stations (whose exposures vary from city centres to lighthouses in coastal waters) to be compared on a mutually consistent basis. It makes use of the fact that the macroscopic pressure gradient field giving rise to the wind will vary only slowly. Shellard states that "The idea is an attractive one but suffers from a number of practical disadvantages." The writer is not entirely in agreement with Shellard concerning the latter part of this remark and feels that some further comments are necessary.

The first disadvantage that Shellard speaks of concerns the fact that in establishing the gradient wind map of the British Isles (which was originally only intended to be illustrative rather than definitive), a subjective method was used to evaluate the roughness at the anemometer station from published descriptions and comparison with sites of known roughness. This is a criticism of the method used and not the principle behind it. Nevertheless, surely even a subjective evaluation of the roughness is better than none at all?

In fact there are a number of methods whereby a non-subjective estimate of the roughness could be made. Most of these methods, however, (measurements of Reynold's stress, roughness length, wind-velocity gradient etc.) could only be carried out effectively by the operators of the stations which, presumably in the United Kingdom is the Meteorological Office itself. The writer is well aware of the imperfections of a subjective evaluation and in fact, at a later stage, a non-subjective method was applied by the writer to evaluate roughness. This depended on using the ratio of gust to average wind speed as an indicator of turbulent intensity and hence also the surface roughness or specifically the surface drag coefficient. The results tended to confirm the earlier subjective estimates but did not seem to offer any real improvement. Unfortunately the method depended on knowing the dynamic response of the meteorological instruments in use about which there is considerable uncertainty and variability.

Perhaps Shellard could be invited to comment on whether or not he feels it would be a profitable undertaking for national meteorological organizations (such as the meteorological office) to investigate the effective surface roughness characteristic and wind profiles at operational meteorological stations.

The second disadvantage Shellard asserts is that the gradient wind approach leads to results which are not consistent with actual measurements. On the whole this criticism appears to be based on different interpretations of observed data. In *fig.5* Shellard purports to show that whereas the gradient wind approach would indicate a range of wind speeds at 33 ft over a variety of different surfaces of some 47 miles per hour that in fact a range of only 21 miles per hour at this height has been observed. This seems misleading.

The so-called observed values are not actual observations but adjusted observations; they do not refer to actual heights but to so-called effective heights. The term effective height of an anemometer is defined by Shellard as "the height at which the anemometer would register a speed equal to that actually observed if there were no obstructions in the vicinity". While one can sympathize with the intention, this concept is, in fact, highly unsatisfactory from the physical point of view and bears little or no relation to boundary layer properties. It cannot be measured because the obstructions in question can seldom be moved; it is not clear whether the term "obstructions" refers also to those elements constituting the statistical roughness of the boundary or not; its value is assessed subjectively, and is highly indeterminate. While it may be a convenient concept for many meteorological purposes, the effective height seems unsuitable for structural engineering applications.

In the accompanying figure the actual observed values of the once-in-50-year-wind (kindly made available by Shellard) are plotted against actual heights of observation. On this basis the agreement between the suggested wind velocity profiles for different roughnesses and the observed velocities is considerably improved.

The same confusion between effective and actual heights seems to lie at the basis of Shellard's preference for the power law exponent of .23 for city areas and his opinion that an index of .40 "would be quite inappropriate". If actual heights are used then there is a growing body of reliable evidence to show that an exponent in the vicinity of .35 - .45 definitely is appropriate over extremely rough surfaces such as a city centre. The well established value of .28 for the wooded terrain at Brookhaven is another clear indication of the trend.

Finally the writer has some sympathy for Shellard's conclusion that the gradient wind approach "cannot give satisfactory results unless much more reliable information becomes available about the values of α and Z_G that are appropriate to different types of terrain". If the phrase "completely satisfactory results" had been used, the writer would be in full agreement. Unfortunately, structural engineers need the best answer they can get now. The questions that really need to be asked are which approach at present gives the most reliable estimate of wind velocity at the site of a structure? Also along what lines could improvements be made?

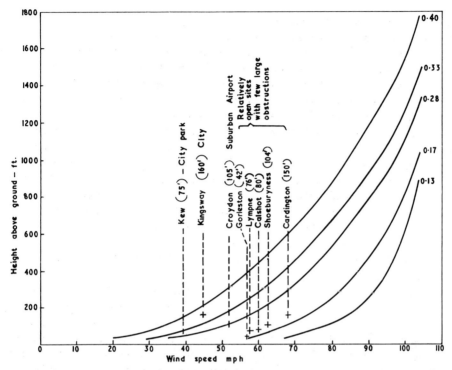

Unadjusted estimates of once–in–50 year hourly wind speeds at instrument height for stns. in S.W. England. Suggested wind velocity profiles shown for comparison.

MR. SHELLARD (in reply). I am sorry that Professor Davenport is disappointed by my reaction to his gradient wind approach. As I tried to indicate in my paper, my reservations are based not on any doubts concerning the soundness of the method, but on doubts about the accuracy of the results obtainable compared with those given by the more conventional approach described in my paper. The accuracy of the latter depends primarily, of course, on the existence of a relatively close network of wind-recording stations, such as we have in the United Kingdom.

The quantitative evaluation of the roughness at a particular site is certainly difficult. Even if measurements are available they may be difficult to interpret. The figures given on page 41 of my paper, relating to Prestwick airport, show that one measure of surface roughness, the ratio of the maximum gust to the mean speed in strong winds, can vary over quite a wide range even at a very open site. Similar results have been obtained for other airfields and this presumably indicates that the effective roughness is appreciably different for different wind directions.

Thus it is not easy to comment on the profitability of making measurements of effective surface roughness and wind profiles at operational meteorological stations. From the point of view of the engineer concerned only with the sites for which he is required to design structures, it would appear to be rather small. Even from the research aspect, it is suggested that it would be less expensive and of greater value to confine any such measurements to a few selected typical sites.

It must be admitted that *fig.5* of my paper is misleading to the extent that the values referred to as observed values are in fact values which have been adjusted to the standard height of 33 feet above the ground, using the effective-height concept. Nevertheless, as Davenport's diagram shows, even the unadjusted observed values cover an appreciably smaller range of speeds than would be obtained by the application of the gradient wind approach, using the profiles seemingly most appropriate to the sites concerned. For example, Kingsway is windier than the 'city' profile would indicate and Gorleston is less windy than the 'flat coast' profile would suggest.

'Effective height' was defined and introduced in the Meteorological Office as long ago as 1932. I must therefore disclaim any personal responsibility for introducing it. As Davenport says, it is a convenient concept for meteorological purposes, but I think that there is rather more to it than this. It is an attempt to take account of the fact that obstacles create eddies which make wind measurements in their vicinity unrepresentative. Since a building affects the wind, not only in its immediate wake but also at levels up to at least twice its height and at distances downwind up to at least twenty times its height, this seems to be a matter of some importance to engineers too. Davenport bases his city maximum wind profile on heights measured from ground level, thus ignoring the fact that the low-level measurements may be unduly influenced by the nearest parts of individual roughness elements (buildings). In a city with buildings ranging in height between, say, 30 and 100 feet, it would be quite possible to have several anemometers all at the same height, say, 130 feet above the ground, indicating appreciably different speeds. Also, the wind-tunnel measurements of Jensen and Franck (see paper 15 and also their paper "Model Tests in Turbulent Wind, Part I" cited therein), suggest that the wind profile over a city is such that heights should be measured, not from ground level but from some higher level; this they found to be at approximately half the height of the buildings in a model representing a densely built-up area of five-storey buildings.

In reply to Davenport's penultimate question, it is suggested that in a country like the United Kingdom, where we have a close network of anemograph stations, the conventional approach is still likely to give the most reliable estimates of maximum wind speed at the site of a structure, especially if a short series of surface wind observations can be made at the site concerned. In a country where the network is more open, but where it is sufficient to permit determination of the maximum gradient wind field, it is possible that the gradient wind approach would be preferable. In any case the gradient wind approach can provide a valuable check and Professor Davenport has undoubtedly made a notable contribution in developing it.

PAPER 13

THE MEASUREMENT OF WIND PRESSURES ON TALL BUILDINGS

by

C. W. NEWBERRY
(Building Research Station)

THE MEASUREMENT OF WIND PRESSURES ON TALL BUILDINGS

by

C. W. NEWBERRY, B.Sc., A.M.I.Mech.E., F.R.P.S.
(Building Research Station)

SUMMARY

THE paper reviews some previous work on the measurement of wind pressures on the full scale and touches on the need for further research. It deals with some of the problems that arise in this work.

After outlining the scope of the current Building Research Station programme on the measurement of wind effects, the paper describes the pilot investigation being made at State House, Holborn, to develop suitable techniques and to study the effects of gusts on the pressure distribution.

1. INTRODUCTION

Builders have always had to contend with wind forces, and it is only natural that through the centuries, by a process of trial and error, the traditional constructions that have been evolved have generally been reasonably resistant to the effects of the wind. The safety factor has, however, often been only marginal and on occasions quite inadequate to meet abnormal wind storms, and recent experience indicates that there is a need for a re-assessment of current building practice to provide for the wind loads that are imposed on structures from time to time.

There have been, moreover, considerable changes in building techniques in recent years which have tended to make buildings more susceptible to the effects of wind. There is an increasing tendency to build high with the result that more buildings are exposed to higher wind speeds, and wind loadings are now assuming a greater significance in relation to the other forces imposed on the building. In addition to this, the use of lighter constructions and lightweight cladding units has made many modern buildings more responsive to short duration gusts, the effects of which are not yet fully understood, but which clearly involve the operation of forces greater than have hitherto been assumed to act. These features, together with the evidence of a number of structural failures due to wind action, have prompted a further investigation by the Building Research Station of the forces acting on buildings during strong winds.

The investigation of wind loads on structures has progressed, spasmodically, over at least eighty years. It was in 1884 that Baker[1] described experiments that he and Fowler had carried out to measure wind loads in connection with the building of the Forth Bridge; and then in 1925 came Stanton's[2] work on the Tower Bridge. In both of these investigations there was an objective to measure the pressures on various parts of the structure and, additionally, to determine the difference of pressure as between a localized point on the structure and a more extensive section of it, i.e. to investigate the effect of the gustiness of the natural wind. Both Baker and Stanton made advances in the knowledge of wind action on structures but the instrumentation available at the time was inadequate to deal with the complexity of the problems to be solved; nevertheless, these early experiments brought an awareness of the great field still to be explored.

One of the first attempts to measure the pattern of wind loading on a tall building was that carried out on the Empire State Building during 1932-36 and reported by Rathbun[3]. Pressure readings were obtained from a series of manometers connected by long pipes to pressure holes on the face of the building. The frequency response of such a system was inevitably too low for the gust pattern to be fully indicated, but the experiment gave a survey of the general pressure distribution, and this was compared with the pressure pattern obtained from a model of the same building tested in a wind tunnel[4]. It was concluded 'that the natural wind movements are not at all like those in a wind tunnel': but one is left with some doubt as to the validity of the pressure pattern obtained since the readings from different parts of the building were not simultaneous, and variation in the incident wind during the period of a set of readings seems to have been very probable.

At the present time there is need for much more data on wind loading to provide a rational basis for structural design. Most of the meteorological records available are of limited value to the structural engineer because they have been obtained largely from measurements in relatively open situations, such as airfields, which are not representative of the sites that generally interest the builder; and, moreover, they do not indicate directly the parameters of the wind that are the concern of the designer. It is widely recognized now that buildings can be damaged by wind pressures of comparatively short duration. If the wind force exceeds the strength of the building to withstand it, the occurrence and extent of damage will depend on the magnitude and duration of the force, the inertia of the building, and the response frequencies of the building and its several parts. The force durations necessary are seldom more than a few seconds, except in the case of long flexible structures such as suspension bridges, and may well be less than a second in the case of lightweight structural components.

Thus it is necessary to measure in detail the actual forces imposed locally on a structure, and to determine the overall loadings, on a time base appropriate to the structure, and to relate these in the most useful manner to the meteorological data currently available. The experimental work to be described was arranged to provide some of these data now urgently required by the designers of tall buildings: the wind speed at various heights over built-up areas: the characteristics of the gustiness: and the effects of these in producing pressures and suctions on the elements of structure.

2. PROBLEMS OF FULL-SCALE MEASUREMENT OF WIND PRESSURES

From the structural point of view the ideal way to determine wind effects on buildings would be to measure the total instantaneous load on complete units of cladding, and on the structure as a whole. Unfortunately this is generally impracticable and it is necessary to compromise by taking sample loadings at discrete points. This can be achieved by fitting wind-pressure gauges flush with the surface of selected buildings, the number and disposition of the gauges being controlled by the nature of the building and the particular programme appropriate to it. As suggested above, the gauges and the whole of the recording equipment should have a sufficiently high response frequency to be able to deal with pressure transients lasting only a fraction of a second (a suitable target is a time resolution to about 0.1 sec.); and the gauges must record continuously and simultaneously on to a common chart in order that the fluctuating pressure pattern on the building may be studied. These requirements virtually necessitate the use of pressure gauges having an electrical output.

A study of meteorological records covering wind and atmospheric pressure and temperature reveals a range of serious problems that must be overcome in any attempt to measure wind pressures on a building. The first, and most important of these is concerned with the range of pressures to be measured and the establishment of a reference pressure. It will be realized that, in any programme of wind-pressure measurement, the majority of readings will be in the range ± 10 lb/ft^2 and that seldom, even in severe storms, will pressures measured on the cladding exceed 30 lb/ft^2 relative to the barometric pressure. Yet, during the period of recording, the barometric pressure itself may change by as much as 200 lb/ft^2; and even during relatively short periods of recording the change of barometric pressure may exceed the pressure variations due to the wind. It is fortunate that barometric changes are generally sufficiently slow for external and internal pressures on a building to become equalized. This large variation of barometric pressure makes it preferable to record wind pressures as increments on the barometric rather than as absolute pressures; that is, changes in the barometer should not affect the readings of the wind pressure gauges. Thus, the pressure capsules employed should not be sealed. They should be open at the back to the atmosphere, but this poses yet another problem: what reference pressure should be adopted? The pressure inside a building or a room will depend on the wind flow around the building and on the amount and distribution of door and window openings, and it may vary from place to place in the building. A gauge directly open at the back will indicate the difference between the external and internal pressures at that point on the building, i.e. the total loading imposed locally on the wall. In some circumstances, this may be the indication required: but, if there are significant pressure differences between different parts of the interior of the building, there will be no common datum for the several gauges in use and it will not be possible to establish the true external pressure pattern or to determine the total external load on the structure. For this it is essential to have a common reference pressure for all gauges. This is being achieved in some of the Building Research Station's installations by running an air pipe from each gauge to a common reservoir which is open to atmosphere but installed in a situation protected as far as possible from external influences other than the normal excursions of the barometer. One necessary precaution in arranging such a system is to minimize the stack effect in any vertical run of pipe. In a warm building this could amount to about 0.5 lb/ft^2 for each 100 ft of height. A solution (in northern temperate regions) is to run the vertical pipe connections externally on the north face of the building where they are exposed to atmospheric temperature but screened from solar radiation.

It will be appreciated that the installation of wind-pressure recording equipment is a complex process and will need to be tailored to the particular building to be studied. For this reason it is desirable to plan the installation during the design stages of the building, and to build in the cables and pipe work in phase with the construction and to install the gauges or their mounts as the cladding progresses. A drawback to such a process is the time taken in building construction. For any major building it is likely to be at least two years after planning the experiment before any wind-pressure measurements can be made. Nevertheless, the advantages of such a procedure appear to justify the method for a major experiment, and most of the Building Research Station's work on wind-pressure measurement has been planned on these lines. An alternative method, which has some disadvantages, but which has been used as a matter of expediency, is to utilize an existing building and to mount wind-pressure gauges in the windows, with connections run as may be convenient.

One other major problem that must be referred to is that of correlating the wind-pressure measurements with the wind conditions prevailing at the time. It is not very probable that pressure measurements will be actually in progress at the very time when maximum loading of the building occurs. Measurements will be made under a variety of wind conditions and must be related to those conditions. Then by extrapolation, maximum probable loadings during the life of the building will be determined by reference to the maximum probable wind speeds estimated from the long term meteorological observations most appropriate to the site. But the measurement of wind speeds incident on a large building presents great difficulties. Ideally an anemometer is required to windward, separated from the building by a sufficient distance so that it will be out of the influence of the building itself: for large buildings tend to deflect and modify the wind flow for an appreciable distance around themselves, in every direction. To meet the conditions of winds from various directions it would be necessary to erect an array of anemometers, but this is usually impracticable. Often the most promising solution is to mount an anemometer on a mast on the roof of the building; but a tall mast is required to reach clear of the local disturbances, and this is not always acceptable to architects and town-planning authorities. Moreover, the fact must be taken into consideration that wind speed and turbulence varies with height and that the anemometer must not be unduly remote from the levels where pressure measurements are to be made. It seems that there can be no complete and perfect solution to this problem of correlating wind speeds and their associated effects and extrapolating them to the probable maximum. Each case must be treated on its merits and the best compromise solution worked out.

3. THE SCOPE OF THE CURRENT TESTS

The review of the previous work on wind loading which had been a necessary preliminary to the present investigation had revealed the need for further research in a number of directions, particularly in the field of full-scale experiment needed to check the validity of wind-tunnel experiments on models. A consideration of the scope of even this single aspect of the subject showed that only a very limited range of experiment was possible, and it was decided to restrict the investigation in the first instance to the problem of tall buildings, because of the considerable interest and activity in this type of building at the present time and the paucity of information relating to it.

Search was made for suitable projects at the design stage and, with the assistance and co-operation of the architects and engineers concerned, two buildings of widely different character were selected, each to provide a part of the information being sought. The first of these buildings is representative of much that is under construction at the present time. It is of plain "match-box" form with a relatively smooth exterior surface and no excrescences that might unduly complicate the wind flow. It is 220 feet high, with a plan area of 140 feet by 58 feet. Known as Block F it is on the Barbican site in the City of London. It is relatively unobstructed by tall buildings to the west but is in the turbulent wake of other similar buildings to the east and north-east, and it will thus be possible to compare the wind-pressure pattern under differing conditions of flow. Block F is being instrumented with 48 pressure measuring points. They are installed on the 7th, 13th and 17th floors at heights of 86 feet, 152 feet and 196 feet above ground level, in an asymmetric pattern designed to utilize different wind directions to complement one another in building up the maximum yield of information on the pressure distribution. This array of gauges, selected groups of which feed their outputs simultaneously on to a multi-channel galvanometer recorder having a chart speed of about 3 inches/minute, will enable the pressure pattern to be studied in considerable detail to provide information on the extent and duration of gust action on the cladding.

The other building selected is the 580 feet high G.P.O. Tower in course of erection at the Museum Telephone Exchange in London. Use of this tower will allow measurements to be made over the greatest possible range of heights. It will go a long way to provide the urgently needed information on the vertical profile of the wind over London. The tower is circular in section and for the most part has an external diameter of 50 feet. This plan form of the tower is admirably suited for a test programme of this nature because it has no bias as regards wind direction: moreover the pressure distribution round a circular tower has been well established in the smooth flow of the wind tunnel and is available for comparison with the measurements in the natural wind. For this reason gauges are to be installed at 12 equally spaced positions round the tower at each of four levels, i.e. at approximately 155 feet, 215 feet, 335 feet and 505 feet above ground level. Additional gauges are being installed at intermediate levels on that part of the surface of the tower facing the prevailing SW winds in order to have as complete a vertical profile as possible of the wind from that quarter. These gauges should also yield information on the vertical extent of gusts, and the manner in which gusts affect a tall building.

Strain gauges have been fitted to the steel reinforcement in the concrete of the tower so that it will be possible to determine by calculation the total wind loading, and to compare this with the results obtained from the measurement of the wind pressure with the individual gauges.

At the tower site it has been possible to arrange for the measurement of wind speeds for direct comparison with the pressure measurements. The tower itself is to be surmounted by a 40 feet lattice mast which will carry an anemometer at a height of about 620 feet; and in addition there is available in the vicinity another lattice tower which will permit the mounting of anemometers at heights of about 250 feet and 130 feet with open exposure to all winds except those from the north.

As mentioned earlier, the installation of wind measuring equipment in buildings under construction is a long-drawn operation governed by the progress of building, and measurements will not be possible before the Spring of 1963 at Barbican and perhaps a year later at the G.P.O. tower. To provide some limited information at an earlier date, arrangements were made to instrument two existing buildings which, though not ideal for purposes of wind measurement, offered facilities that were of immediate value. One of these, the Millbank Tower, is the tallest building in London at the present time with a height of 387 feet. It is being equipped with wind-pressure gauges set flush with the walls near the centre lines of the west and south faces at heights of 80 feet, 150 feet, 220 feet, 275 feet and 325 feet and will be in operation shortly, to give some preliminary indication of the pressure characteristics over this height range.

The other building which has been instrumented for wind-pressure measurement is State House, in Holborn, London, which houses the Headquarters of D.S.I.R. State House is far from ideal for a general study of wind-pressure distribution because it has a complex plan form and has a series of external frames which interfere with the wind flow on the longer faces of the building. It had, however, the great advantage of being

readily available and was considered to be suitable for a preliminary
experiment. It has already yielded some interesting information on wind
effects, details of which are given in the following section.

4. WIND-PRESSURE MEASUREMENTS AT STATE HOUSE

State House is a 15-storey office building having a plan form as shown
in *fig.1*, with a wing of 9-storey height as indicated. It became available in August, 1961; and, because of its accessibility and the fact that
it was the first building to be fitted with the equipment developed by the
Building Research Station for this detailed study of wind pressures, the
installation was regarded as a prototype to examine some of the characteristics of gust action and to give experience in the techniques of high-
resolution wind-pressure recording.

Wind-pressure gauges were mounted in windows on the east and west faces
of the south block at the 11th and the top storeys as indicated in *fig.1*.
These faces were selected because they are free of obstructions such as
encumber the other faces of the building (*fig.2*). The 11th and 15th
storeys were chosen for instrumentation because the 11th is the lowest that
is clear above the surrounding buildings, and the 15th because it gives the
maximum possible separation from the 11th - a feature which appeared to be
of value in this pilot study of gust action. It was realized that the top
storey might prove to be too near the top edge of the building and that
severe falling-off of the pressure coefficient might be found at this level,
but this risk was accepted, and the decision has proved to have been justified, although, as will be shown by a survey of pressures inside the
building by means of an aneroid barometer, pressures at the 15th floor are
somewhat lower than at the 14th.

Four additional gauges were mounted on the south and north faces, as
indicated in *fig.1*, not so much to study the pressures at these positions
as to serve as intermediate markers to help identify gusts on their passage
from end to end of the building.

The wind-pressure gauges are described in appendix A. They were
mounted directly into windows as shown in *fig.3* and were connected by
electrical cables to recording equipment installed in a room on the 15th
floor. Because of the impracticability in this case of connecting all
gauges to a common reference pressure, each gauge was made open at the back
and thus relied on the pressure of the room in which it was situated as its
datum.

There was, at the beginning of this experiment, no information, so far as
could be discovered, about the pressure variations to be expected in any

one room of the building nor of the pressure differences to be expected between different rooms. A preliminary survey was therefore made to investigate the scale of such possible variations since they might have an important bearing on the evaluation of the external pressures that would be indicated by the gauges in the windows. The survey was made by means of a sensitive aneroid barometer, readings of which were noted in various parts of the building, under various conditions of door and window openings, and on different occasions with different conditions of wind. In addition to noting the height of the barometer above ground level the time of each reading was noted so that allowance could be made for the variation of the barometric pressure with time.

It was found that, with a strong wind blowing, windows on the windward face of the building were generally closed. Under these conditions, pressure throughout the building was reasonably uniform (after allowing for height above ground level) and approximated to the external pressure on the leeward face. If, however, a window were open on the windward face and the door of the room closed the pressure in that room was subject to rapid fluctuation due to gust action, and the range of pressures was surprisingly large, being comparable with the probable range of external pressures. For example, in a westerly wind gusting generally to 41 m.p.h. with a maximum gust of 44 m.p.h. the corresponding dynamic pressure heads q being 4.2 lb/ft^2 and 4.8 lb/ft^2 respectively, typical pressures in west-facing rooms with windows partly open were:

on the 14th floor, 1.6 lb/ft^2 rising to 4.2 lb/ft^2 in gusts;
on the 15th floor, 1 lb/ft^2 rising to 2.8 lb/ft^2 in gusts;

all of these being relative to the prevailing pressure in the corridors at each floor level, which was, as mentioned above, approximately that of the leeward face of the building. Some windward rooms showed pressure surges of the order of 0.5 lb/ft^2 when the windows were nominally closed. This suggests that in some cases there may be errors of up to about 10% in the measured pressures on windward faces when the room pressure is taken as the datum, even though the windows are closed; and it emphasizes the advantage to be gained by using a common reference pressure for all gauges. Some details of the survey are given in appendix B.

Wind-pressure records were taken from time to time during 1962 under a variety of wind conditions. The installation was regarded as experimental and subject to development, and for the first part of the year the control was manual, the equipment being switched on when conditions were deemed favourable. It so happened that in the early part of the year there were several severe gales, but they mostly blew up with insufficient warning for the pressure recorder to be manned, and the records so far obtained cover only moderately strong winds. Automatic control gear has since been built and installed, and at the present time the equipment samples at regular intervals, for example, 10 minutes out of each hour, with provision for an over-riding control from an anemometer to continue to record if the wind speed exceeds a pre-determined value. Specimen records are shown in *figs.4, 5, 6* and *7*. It will be noticed that in each case 12 channels have been recorded, appropriate gauges having been selected to examine various features of the wind effect. It had originally been intended to record 24 channels on the chart, but such is the erratic nature of wind pressure, and so variable the trace amplitude, that it proved impracticable to crowd the traces. The chart speed was about 3 inches/min., the time base being at 1 sec. intervals in *fig.4*, and at 2 sec. intervals on the other charts.

Turning first to *fig.4*, which includes the whole array of gauges on the east face of the building, it is seen that there is a remarkable correspondence between the pressure patterns of all these gauges, down to the very fine details of gusts lasting only a second or two. The synchronization of the gust action over the area explored (40 feet high by 50 feet wide) under the action of an oblique, south-easterly wind is also quite striking. As will be seen at time 55, and also at 173 and 277, the lag in pressure rise between the gauges at 15E5 and 15E1 is about 1 sec. In the case of the first and second of these gusts there is no discernible lag between the 15th and 11th floors, while in the case of the gust at 277 the pressure rise occurred at the 15th floor about 1.1/2 secs. before it reached the 11th floor. Another feature which should be noted at this stage, and will be seen in most of the records, is the rapidity of pressure fluctuation; peak values being reached from near zero in less than 1 sec. on many occasions, and even reversals of loading from pressure to suction in similar time intervals.

There were few occasions when records could be obtained of winds blowing normal to one of the instrumented faces of the building. One of these records, though with winds gusting only to 25 m.p.h., is shown in *fig.5*. This record, as do the next two reproduced, covers gauges on both the east and west faces of the building, at 11th floor level, with two additional gauges added in each case. Gauges on the windward face indicate pressures that are consistently positive with the exception of 11W1, which is at the trailing edge of the face (the wind was not quite normal to the face). On the leeward face pressures are negligible on gauges 11E2 and 11E3, but the gauges towards the corners of the building indicate suctions at those positions. 15E3, at the centre of the leeward face, top floor, also indicates suction suggesting that negative pressures are operative all round the perimeter of the leeward face. A more complete examination of the effects with winds normal to the faces must await the occurrence of stronger winds from the appropriate directions.

Fig.6 shows a record with winds gusting generally up to 37 m.p.h. from the NE. The close similarity between gauges across the east face is seen again. Also well marked on this record is the variation of pressure intensity across the face, the pressures reaching a maximum at position 11E2, the second gauge from the leading edge, and falling to a minimum at 11E5, which is at the trailing edge of this face. The gust recorded at time 26 shows a progression across the east face in about 1 sec., but in general on this record the gust action is closely synchronized across the face at 11th floor level. This is particularly well shown at time 544 where a sudden pressure rise occurs simultaneously over the whole face. It is to be noted on this record that the gauges on the west, the leeward, face show very little variation of pressure and that the pressures are generally near zero (referred, of course, to the pressure inside the building). It is of interest that some positive surges were recorded on the leeward face, a feature that was later confirmed by watching fans that are installed in the windows of some rooms. These fans were specially arranged to be closeable to eliminate their effect on room pressures during particular tests, but, when open and not switched on, they were seen to reverse direction from time to time under the influence of the external pressure, on the leeward face. The effect of a southerly wind, that is, a wind blowing approximately parallel to the two instrumented faces of the building, is shown in *figs.7a* and *7b*. There is little in common in the pressure patterns of the two ends of the building except that both show a maximum of wind effect at the windward edges of the faces, and both show a preponderance of suction at these positions. The west face shows more wind activity suggesting that the wind direction was slightly from that quarter rather than square to the building and, in conformity with this, significant positive pressures occur as transients on the west face whilst the east face remains almost entirely under the influence of suction.

The important feature to notice is the rapid and substantial pressure change from positive to negative and vice versa, particularly at position 11W1, and also at 11W2, and to a lesser extent on the other gauges on the west face. Many of the suction peaks, whilst substantial in intensity, are of very short duration and appear to cover not more than about three gauge positions simultaneously, that is, less than 30 feet width of the building face.

Since the automatic control equipment for the wind pressure recorder was put into operation in December 1962 the wind has been, up to the time of preparation of this paper, predominantly from the NE, which, although not a preferred direction as far as measurements at State House are concerned, has provided a long series of comparable records which have been examined in some detail. The purpose of this has been to explore the wind effects and to investigate methods of presenting them. This information is not generally applicable since, as stated earlier, State House is not a typical form of building. Some of the results together with some relating to a southerly wind, are presented in appendix C. The records were measured to determine the normal range of pressure peaks at each gauge position, and the extreme range. The latter is of some interest, but the former is regarded as a more useful parameter since it can be more readily correlated with the records of wind speed and direction. There can be no certainty that the extreme wind speeds were the same at State House and at the anemometer site which is on the Meteorological Office in Kingsway, about 1/4 mile away, though it seems reasonable to assume that the general level of gust speeds was similar, the anemometer at Kingsway being at about the level of the 15th floor of State House. Mean levels of the pressure at each gauge position for the duration of the record were also derived, but while these may be significant in the case of winds that are normal to the face of the building, it was found that they had little meaning when the wind direction was on to a corner. This was because swinging of the wind direction during gusts could cause alternate positive and negative pressure regions which, although significant in themselves, could have a mean approaching zero.

The mean pressure levels of each record were correlated with the dynamic pressure head of the corresponding wind speed measured at Kingsway, and a coefficient was derived expressing the pressure as a proportion of the dynamic head. This was done also for the normal range of gust pressures, relating them to the pressure heads of the normal gust speeds, and similarly for the extreme gusts. The results, which are given in appendix C, are summarized in *Tables 1* and *2*.

It is to be noted that, if the mean wind speed over the few minutes duration of the record had been taken as the basis of assessment and the pressures recorded during the maximum gusts had been related to the dynamic pressure heads of the mean wind speeds, the pressure coefficients would have been much higher, reaching +2.2 to - 1.2 for the east face under the action of NE winds and +1.4 to -2.3 for the west face under a S wind.

It will be seen that there are considerable variations in the pressure coefficients from one record to another, particularly in *Table 1*. Individual discrepancies are likely to be due to a lack of precise correlation between the recorded pressure and the recorded wind speed, which is a feature that it is intended to improve as far as possible in future experiments. While successive records from the same wind direction show similar trends of pressure distribution, the means of the several records show a marked regularity of pattern, and may well be a useful indication of the relative pressures around the building. It is of interest that the

Table 1

Pressure coefficients for NE winds

(a) Coefficients based on normal gust speeds at Kingsway

Time of Record	18.1.63 17.25		18.1.63 19.25		18.1.63 20.25		19.1.63 12.25		19.1.63 13.25		Mean	
Gauge position												
15E3	+.85	0	+.45	0	+.6	-.05	+.4	-.1	+.55	0	+.6	-.05
11E1	+.65	-.3	+.3	-.2	+.4	-.5	+.4	-.3	+.4	-.3	+.4	-.3
11E2	+.75	-.3	+.4	-.2	+.6	-.3	+.35	-.25	+.5	-.3	+.55	-.25
11E3	+.85	0	+.35	0	+.6	0	+.25	-.15	+.5	0	+.5	-.05
11E4	+.45	-.1	+.3	-.05	+.35	-.15	+.2	-.15	+.3	-.1	+.3	-.1
11E5	+.35	0	+.2	-.05	+.3	-.15	+.15	-.05	+.3	0	+.25	-.05
11W1	+.1	-.1	+.05	-.05	0	-.15	0	-.05	0	-.1	+.05	-.1
11W2	+.1	-.1	0	-.05	0	-.1	0	-.05	0	-.05	0	-.05
11W3	+.2	-.2	+.1	-.1	0	-.2	+.05	-.05	+.05	-.2	+.1	-.15
11W4	+.1	-.1	0	-.05	0	-.15	+.05	0	0	-.15	+.05	-.1
11W5	+.1	-.1	+.05	0	+.1	-.05	+.05	0	+.05	-.05	+.1	-.05
15NW	+.1	-.15	+.05	-.15	+.1	-.15	+.05	0	+.05	-.25	+.1	-.15

(b) Coefficients based on extreme gust speeds

15E3	+.75	-.2	+.4	-.05	+.6	-.15	+.7	-.35	+.6	-.3	+.6	-.2
11E1	+.7	-.25	+.4	-.2	+.6	-.6	+.65	-.3	+.6	-.3	+.6	-.35
11E2	+.75	-.35	+.5	-.4	+.7	-.35	+.75	-.45	+.6	-.3	+.65	-.35
11E3	+.75	-.15	+.45	-.1	+.7	-.15	+.55	-.35	+.55	-.2	+.6	-.2
11E4	+.4	-.1	+.25	-.05	+.3	-.2	+.4	-.25	+.35	-.15	+.35	-.15
11E5	+.4	-.05	+.15	-.1	+.3	-.25	+.25	-.1	+.3	-.05	+.3	-.1
11W1	+.1	-.1	+.05	-.05	0	-.2	+.05	-.05	+.05	-.05	+.05	-.1
11W2	+.1	-.1	+.05	-.05	0	-.1	+.05	-.05	+.05	-.05	+.05	-.05
11W3	+.2	-.2	+.15	-.1	0	-.2	+.1	-.1	+.05	-.2	+.1	-.15
11W4	+.2	-.1	+.05	-.05	+.05	-.1	+.05	0	+.05	-.1	+.1	-.05
11W5	+.1	-.1	+.05	-.05	+.1	-.05	+.05	0	+.05	-.1	+.05	-.05
15NW	+.1	-.25	+.05	-.1	+.1	-.15	+.1	-.4	+.05	-.3	+.1	-.25

Table 2

Pressure coefficients for S winds

(based on corresponding wind speeds at Kingsway)

	Normal Gusts						Extreme Gusts					
	9.3.63 13.35		9.3.63 15.35		Mean		9.3.63 13.35		9.3.63 15.35		Mean	
15E3	0	-.35	0	-.25	0	-.3	0	-.3	0	-.35	0	-.35
11E1	+.05	-.3	0	-.2	+.05	-.25	+.1	-.3	0	-.2	+.05	-.25
11E2	0	-.3	0	-.25	0	-.25	+.1	-.3	0	-.25	+.05	-.3
11E3	0	-.2	0	-.15	0	-.15	+.05	-.2	+.05	-.2	+.05	-.2
11E4	+.05	-.3	+.05	-.2	+.05	-.25	+.1	-.3	+.1	-.2	+.1	-.25
11E5	+.1	-.35	+.05	-.2	+.1	-.3	+.1	-.3	+.1	-.25	+.1	-.3
11W1	+.05	-.55	+.1	-.5	+.1	-.55	+.05	-.55	+.4	-.55	+.25	-.55
11W2	+.15	-.6	+.3	-.4	+.2	-.5	+.3	-.6	+.4	-.4	+.35	-.5
11W3	+.2	-.35	+.2	-.25	+.2	-.3	+.25	-.35	+.3	-.2	+.3	-.3
11W4	+.1	-.15	+.2	-.1	+.15	-.15	+.1	-.15	+.25	-.1	+.2	-.15
11W5	+.1	-.15	+.1	-.1	+.1	-.15	+.1	-.15	+.15	-.1	+.15	-.15
15W4	+.05	-.1	+.05	-.05	+.05	-.1	+.1	-.15	+.05	-.1	+.1	-.15

pattern has a general similarity to, but some differences from, the patterns obtained in wind-tunnel experiments on a model of comparable shape[5]. In view, however, of the very low values of the coefficients realized on the west, the leeward face, during NE winds, there must still be doubt about the suitability of the room pressure as a datum, and a possible error due to this cause must be borne in mind in attempting to utilize these results.

Another feature of the wind effect that has been abstracted from some of the records of NE winds analysed in appendix C is a relationship between the peak pressure exerted at a point on the structure and the time for which it acts. Samples of these results are given in *Table 3*, in each case taken from gauge 11E2, which showed a maximum effect under these winds.

Table 3

	Mean Pressure lb/ft^2	
Duration of gust (secs.)	18.1.63 17.25	19.1.63 12.25
	$q(max) = 4.0$	$q(max) = 5.1$
1	2.9	3.6
2	2.8	3.3
3	2.7	3.0
5	2.5	2.8
10	2.2	2.5
60	1.1	1.6
600	0.8	0.45

These pressures rise considerably more steeply as the time interval shortens than do the corresponding figures derived from Durst's[6] gust factors. This supports the hypothesis, which can be deduced from an inspection of the wind-pressure records, that variations of pressure on the surface of a building are due more to variation of the angle at which the wind strikes the building than to variations of the speed of the wind; that is, it is the change in the angle of attack when a gust traverses a building that causes these rapid fluctuations in pressure. This effect needs to be explored further, and under a variety of wind directions and intensities.

CONCLUSIONS

The pilot experiment of wind-pressure recording at State House has indicated some of the problems to be solved before arriving at a fuller understanding of wind action on a building. Of these, the most difficult, yet most important, is the establishment of a suitable reference pressure against which the external pressures can be measured. In the experiments carried out so far at State House, high suction coefficients have been observed only in the case of surges lasting for a very few seconds, but it would be unwise to draw broad conclusions from this observation until an opportunity has come to check this effect at least on another building with a more complete gauge installation used with a more satisfactory pressure datum.

The rapidity of the pressure changes and the way they move across the faces of the building are interesting features. Positive pressures appear to build up rapidly and to act simultaneously over large parts of the surface, while the principal suction peaks, which occur with the wind at a glancing angle to the face of the building, appear to be of very short duration and to act over small parts of the surface only at any one instant. The major short duration pressure changes are linked with changes of wind direction caused by turbulence rather than by changes in the speed of the incident wind. This suggests that a new approach may be desirable in the consideration of gust factors, and that there may be important differences between the effects of the natural wind and the steady conditions used in wind-tunnel testing. A more complete investigation of this will be possible in the course of the experiments being set up at Barbican and the G.P.O. Tower.

Finally, it must be re-iterated that the work at State House is primarily exploratory and as yet incomplete. The major investigation covering tall buildings is yet to come.

ACKNOWLEDGEMENTS

A programme of this magnitude must necessarily be based on teamwork; credit in this work is due to the whole of the Instrument Section at the Building Research Station who, under the guidance of Mr. R. S. Jerrett, have carried out the preparation, installation and operation of the equipment.

Very willing co-operation has been obtained from the Meteorological Office, both in discussion of the problems of wind measurement, and in making available their records of wind at Kingsway.

The work is being carried out as part of the research programme of the Building Research Board of the D.S.I.R. and this paper is published by permission of the Director of Building Research.

REFERENCES

1. BAKER, B., The Forth Bridge. *Engineering* 1884, **38**.
2. STANTON, T. E., Report on the Measurement of the Pressure of the Wind on Structures. *Proc. Inst. civil Engrs.* 1925, **219,** 125.
3. RATHBUN, J. C., Wind Forces on a Tall Building. *Trans. Amer. Soc. civil Engrs.*, 1940, **105.**
4. DRYDEN, H. L. and HILL, G. C., Wind Pressure on a Model of the Empire State Building. *J. Res. nat. Bur. Stand.* 1933, **10.**
5. NING CHIEN, YIN FENG, HUNG-JU WANG, and TIEN-TO SIAO, Wind Tunnel Studies of Pressure Distribution on Elementary Building Forms. *Iowa Inst. of Hydraulic Research*, 1951.
6. DURST, C. S., Wind-Speeds over Short Periods of time. *Meteor. Mag.* 1960, **89.**

APPENDIX A

The Building Research Station's Wind-Pressure Gauge

The Building Research Station's wind-pressure gauge was designed to be mounted flush with the surface of the building to be studied and to have as large a measuring area as practicable. It therefore takes the form of a circular pressure plate, about 4 in. diameter, which is housed within an annular frame. It is supported within the frame by three double cantilever strips, as may be seen in $fig.8$. The cantilevers are flexible in a direction normal to the pressure plate but are stiff transversely so as to keep the pressure plate concentric in the frame. The narrow annular gap between the frame and the pressure plate is closed by a thin membrane of "melinex" to prevent air leakage from the front of the gauge to the inside. The dimensions of the strips are arranged so that the displacement of the pressure plate is 0.005 in. under a loading of 25 lb/ft^2, and is no more than 0.002 in. under the usual working conditions. The pressure plate is made as light as possible consistent with the requirement of adequate rigidity at the cantilever mountings, and, in consequence, the natural frequency of the gauge has been kept sufficiently high, being in excess of 50 cycles/sec.

The load on the wind-pressure gauge is measured by recording the strain in the cantilever strips by means of resistance strain gauges. Four strain gauges are attached to each strip, two on each side, so that each double-cantilever, when it deflects, puts two strain gauges into tension and the other two into compression. These are wired suitably into the four arms of a bridge circuit, and so give the maximum possible sensitivity and at the same time provide complete compensation for thermal strain in the cantilever. All three cantilevers are treated in the same manner and the corresponding strain gauges on each are connected in series in each bridge arm to ensure that each cantilever contributes its share to the output signal according to the load imposed on it.

Originally the cantilevers were made of phosphor-bronze in order to take advantage of the relatively low Young's Modulus of that material and obtain the greatest possible gauge sensitivity; but in the course of development a change was made from wire resistance gauges to foil type gauges because of their smaller physical size and it was found that the Araldite cement used with these foil gauges was unsuitable for bonding on to phosphor-bronze. Some later pressure gauges were therefore made with steel cantilevers, suitably modified in dimensions to give approximately the same sensitivity.

The gauges were calibrated by dead weight, and the calibration was confirmed by air pressure, in both positive and negative directions, and gave an output corresponding to about 40 units of strain per lb/ft^2. This, at a suitable attenuator setting in the amplifiers gives about 4 mm deflection on the recorder chart per lb/ft^2, which is the scale used in the records illustrated in this paper.

The gauges were subjected in the laboratory to extremes of temperature likely to be encountered in service and the effects were found to be negligible. Some difficulties were, however, encountered when the gauges were put into service, some of them showing excessive zero drift. This was traced mainly to creep in the metal of the gauge bodies following the machining operation. It has been much reduced by careful selection of the material of which the gauges are now made, and by appropriate treatment, but it led to a trial of a further variant of the cantilever system which appears to be successful in minimizing the effects of strain in the bodies. This was to substitute simple cantilevers for the double ones, each cantilever being fixed rigidly to the frame of the gauge but having a universal pivot connection to the pressure plate. This improvement is achieved at the cost of some increase in the complexity of the manufacturing process and the method is only used if the rigid mounting proves troublesome.

APPENDIX B

Survey of Internal Pressures in State House

Condition   Strong westerly wind - 8th August, 1961
               typical gusts 41 m.p.h. $q = 4.2$ lb/ft^2
               max. gusts 44 m.p.h. $q = 4.8$ lb/ft^2

Readings were taken with an aneroid barometer at various positions in the building, with various conditions of window and door openings. All barometer stations were at sill height to enable comparisons to be made. Since the barometric pressure was itself rising by 0.085 cm/hour during the course of the survey, all readings have, for the purposes of comparison, been adjusted to a common point in time. They have also been adjusted to a common height above ground, according to the relationship, for a standard atmosphere, that 100 ft of height corresponds to a pressure difference of 0.272 cm of mercury. The correction per storey height, which is 10.1 feet, is thus 0.02745 cm.

The reference point is taken as 15.15 hrs at 14th floor level, and results are tabulated overleaf.

The aneroid used for this survey had a scale marked in divisions of 0.1 cm and it was necessary to judge the second decimal place by eye and to take great care to avoid parallax errors. It will therefore be seen that the instrument was barely adequate to deal with the small variations of pressure within the building, and too much reliance should not be placed on the last digit. There is, however, a fairly general agreement that the interior pressure was 75.32 ± 0.01 (adjusted to the level of the 14th floor) when windows were closed: that it varied by very little when east face windows were opened: but that when windows were opened on the windward side there were variations of pressure amounting to about 0.15 cm, that is 4.2 lb/ft^2, during gusts, at a time when the maximum dynamic pressure head of the measured gusts was 4.8 lb/ft^2.

Time	Location ref. fig. 1	Floor level	Conditions	Readings and adjustments		Adjusted values
14.40	A	Ground	interior space windows closed door open	time correction + height " − " − total	75.63 0.05 0.38 0.33	75.30
14.55	B	15	lift vestibule (ventilated)	time correction + height " + " + total	75.25 0.03 0.03 0.06	75.31
15.00	C	15	by open window east side	time correction + height " + " + total	75.26 0.02 0.03 0.05	75.31
15.05	D	15	west side room window closed door open	time correction + height " + " +	75.28 (.26)* 0.01 (.29) 0.03	75.32 (.30)* (.33)
			window open door closed	total 	0.04 75.31 (.29)* (.37)	75.35 (.33)* (.41)

Time	Location ref. fig.1	Floor level	Conditions	Readings and adjustments	Adjusted values
15.15	E	14	lift vestibule (ventilated)	75.33(steady)	75.33
	F	14	SE corner room window open door closed	75.33(steady)	75.33
15.20	G	14	SW corner room window closed door open	75.32	75.32
			window open door closed	75.36(.35)* (.50)	75.36(.35)* (.50)
	H	14	NW corner room window closed door open	75.33	75.33
			west face window open door closed	75.39(.36)* (.48)	75.39(.36)* (.48)

* range in gusts

APPENDIX C

Analysis of wind pressure records

Record from 18.1.63 : 17.25 hrs.
mean wind direction NE (55°)
range 350° to 90°

mean wind speed	20 m.p.h.	$q = 1.1$ lb/ft^2
frequent gusts to	34 m.p.h.	$q = 2.9$ lb/ft^2
max. gusts to	40 m.p.h.	$q = 4.0$ lb/ft^2

Gauge	Mean wind		Normal gusts				Extreme gusts			
	press. lb/ft^2	coeff.	press. range lb/ft^2		coeff.		press. range lb/ft^2		coeff.	
15E3	1.0	.9	2.5	0	.85	0	2.9	-.7	.75	-.2
11E1	.45	.4	1.9	-.9	.65	-.3	2.7	-1.1	.7	-.25
11E2	.8	.7	2.2	-.9	.75	-.3	3.0	-1.35	.75	-.35
11E3	1.0	.9	2.5	0	.85	0	2.9	-.55	.75	-.15
11E4	.35	.3	1.35	-.2	.45	-.1	1.6	-.45	.4	-.1
11E5	.35	.3	1.1	0	.35	0	1.5	-.2	.4	-.05
11W1	-.1	-.1	.2	-.2	.1	-.1	.35	-.35	.1	-.1
11W2	-.1	-.1	.2	-.35	.1	-.1	.35	-.35	.1	-.1
11W3	-.2	-.2	.55	-.55	.2	-.2	.8	-.8	.2	-.2
11W4	-.2	-.2	.35	-.35	.1	-.1	.7	-.45	.2	-.1
11W5	0	0	.2	-.2	.1	-.1	.35	-.35	.1	-.1
15NW			.2	-.45	.1	-.15	.45	-.10	.1	-.25

Record from 18.1.63 : 19.25 hrs.
Mean wind direction NE (55°)
Range 20° to 90°

mean wind speed	29 m.p.h.	$q = 2.2$ lb/ft^2	
frequent gusts to	37 m.p.h.	$q = 3.6$ lb/ft^2	
max. gusts to	44 m.p.h.	$q = 4.9$ lb/ft^2	

Gauge	Mean wind		Normal gusts				Extreme gusts			
	press. lb/ft^2	coeff.	press. range lb/ft^2		coeff.		press. range lb/ft^2		coeff.	
15E3	.45	.2	1.6	0	.45	0	1.9	-.2	.4	-.05
11E1	.1	.05	1.0	-.7	.3	-.2	1.9	-.9	.4	-.2
11E2	.45	.2	1.5	-.7	.4	-.2	2.4	-2.0	.5	-.4
11E3	.55	.25	1.35	0	.35	0	2.1	-.45	.4	-.1
11E4	.1	.05	1.0	-.2	.3	-.05	1.35	-.35	.25	-.05
11E5	.1	.05	.7	-.2	.2	-.05	.8	-.45	.15	-.1
11W1	-.1	.05	.2	-.2	.05	-.05	.35	-.2	.05	-.05
11W2	-.1	.05	0	-.2	0	-.05	.2	-.2	.05	-.05
11W3	-.1	.05	.45	-.45	.1	-.1	.8	-.45	.15	-.1
11W4	-.1	.05	0	-.2	0	-.05	.35	-.2	.05	-.05
11W5	0	0	.2	0	.05	0	.35	-.2	.05	-.05
15NW	0	0	.2	-.55	.05	-.15	.2	-.55	.05	-.1

Record from 18.1.63 : 20.25 hrs.
Mean wind direction NE (55°)
Range 20° to 95°

mean wind speed	19 m.p.h.	$q = 1.0$ lb/ft^2
frequent gusts to	32 m.p.h.	$q = 2.6$ lb/ft^2
max. gusts to	36 m.p.h.	$q = 3.1$ lb/ft^2

Gauge	Mean wind		Normal gusts				Extreme gusts			
	press. lb/ft^2	coeff.	press. range lb/ft^2		coeff.		press. range lb/ft^2		coeff.	
15E3	.65	.65	1.6	-.1	.6	-.05	1.9	-.45	.6	-.15
11E1	.2	.2	1.0	-1.35	.4	-.5	1.8	-1.8	.6	-.6
11E2	.55	.55	1.6	-.8	.6	-.3	2.2	-1.1	.7	-.35
11E3	.65	.65	1.5	0	.6	0	2.1	-.45	.7	-.15
11E4	.1	.1	.9	-.45	.35	-.15	1.0	-.65	.3	-.2
11E5	.1	.1	.8	-.45	.3	-.15	.9	-.8	.3	-.25
11W1	-.1	-.1	0	-.45	0	-.15	0	-.55	0	-.2
11W2	-.1	-.1	0	-.2	0	-.1	0	-.35	0	-.1
11W3	-.2	-.2	0	-.55	0	-.2	0	-.7	0	-.2
11W4	-.1	-.1	0	-.35	0	-.15	.1	-.35	.05	-.1
11W5	0	0	.2	-.1	.1	-.05	.35	-.2	.1	-.05
15NW	0	0	.2	-.45	.1	-.15	.35	-.45	.1	-.15

Record from 19.1.63 : 12.25 hrs.
Mean wind direction NE (45°)
Range 330° to 100°

mean wind speed	23 m.p.h.	$q = 1.4$ lb/ft^2	
frequent gusts to	42 m.p.h.	$q = 4.4$ lb/ft^2	
max. gusts to	46 m.p.h.	$q = 5.1$ lb/ft^2	

Gauge	Mean wind		Normal gusts				Extreme gusts			
	press. lb/ft^2	coeff.	press. range lb/ft^2		coeff.		press. range lb/ft^2		coeff.	
15E3	.65	.45	1.8	-.45	.4	-.1	3.6	-1.8	.7	-.35
11E1	.2	.15	1.8	-1.35	.4	-.3	3.3	-1.6	.65	-.3
11E2	.45	.3	1.6	-1.1	.35	-.25	3.8	-2.2	.75	-.45
11E3	.35	.25	1.1	-.65	.25	-.15	2.7	-1.8	.55	-.35
11E4	.2	.15	.9	-.55	.2	-.15	2.1	-1.35	.4	-.25
11E5	.2	.15	.55	-.2	.15	-.05	1.35	-.55	.25	-.1
11W1	0	0	.1	-.2	0	-.05	.2	-.35	.05	-.05
11W2	0	0	.1	-.2	0	-.05	.2	-.35	.05	-.05
11W3	.1	.05	.2	-.35	.05	-.05	.65	-.45	.1	-.1
11W4	.1	.05	.35	-.1	.05	0	.35	-.1	.05	0
11W5	.1	.05	.35	0	.05	0	.35	-.1	.05	0
15NW	0	0	.35	-.55	.05	-.15	.45	-2.0	.1	-.4

Record from 19.1.63 : 13.25 hrs.
Mean wind direction NE then SE (40° then 120°)
Range 350° to 90° then 85° to 220° (SW)

mean wind speed	possibly 23 m.p.h. (difficult to assess)	$q = 1.4$ lb/ft^2
frequent gusts to	46 m.p.h.	$q = 5.1$ lb/ft^2
max. gusts to	51 m.p.h.	$q = 6.5$ lb/ft^2

Gauge	Mean wind		Normal gusts				Extreme gusts			
	press. lb/ft^2	coeff.	press. range lb/ft^2		coeff.		press. range lb/ft^2		coeff.	
15E3	1.1	.8	2.9	0	.55	0	3.8	-2.0	.6	-.3
11E1	.1	.1	2.0	-1.6	.4	-.3	3.8	-1.8	.6	-.3
11E2	.8	.55	2.5	-1.6	.5	-.3	3.8	-2.0	.6	-.3
11E3	.8	.55	2.6	0	.5	0	3.6	-1.35	.55	-.2
11E4	.45	.3	1.6	-.45	.3	-.1	2.1	-1.0	.35	-.15
11E5	.45	.3	1.5	0	.3	0	2.0	-.35	.3	-.05
11W1	-.2	-.15	0	-.45	0	-.1	.2	-.55	.05	-.05
11W2	-.1	-.1	.1	-.2	0	-.05	.2	-.35	.05	-.05
11W3	-.35	-.25	.2	-.9	.05	-.2	.35	-1.35	.05	-.2
11W4	-.35	-.25	0	-.7	0	-.15	.2	-.7	.05	-.1
11W5	0	0	.35	-.35	.05	-.05	.35	-.7	.05	-.1
15NW			.2	-1.2	.05	-.25	.45	-2.0	.05	-.3

Record from 9.3.63 : 13.35 hrs.
Mean wind direction S (190°)
Range 120° to 270°
mean wind speed   23 m.p.h. $q = 1.4$ lb/ft^2
frequent gusts to   44 m.p.h. $q = 4.9$ lb/ft^2
max. gusts to   51 m.p.h. $q = 6.5$ lb/ft^2

Gauge	Mean wind		Normal gusts				Extreme gusts			
	press. lb/ft^2	coeff.	press. range lb/ft^2		coeff.		press. range lb/ft^2		coeff.	
15E3	-.8	-.55	-.1	-1.6	0	-.35	0	-2.1	0	-.3
11E1	-.2	-.15	+.2	-1.35	+.05	-.3	+.8	-1.9	+.1	-.3
11E2	-.8	-.55	0	-1.45	0	-.3	+.7	-2.0	+.1	-.3
11E3	-.35	-.25	+.1	-1.0	0	-.2	+.2	-1.25	+.05	-.2
11E4	-.55	-.4	+.2	-1.45	+.05	-.3	+.55	-1.9	+.1	-.3
11E5	-.55	-.4	+.45	-1.8	+.1	-.35	+.8	-2.0	+.1	-.3
11W1	-.9	-.65	+.2	-2.7	+.05	-.55	+.45	-3.6	+.05	-.55
11W2	-.45	-.3	+.7	-2.9	+.15	-.6	+1.9	-3.8	+.3	-.6
11W3	-.2	-.15	+1.0	-1.7	+.2	-.35	+1.7	-2.2	+.25	-.35
11W4	0	0	+.45	-.8	+.1	-.15	+.55	-1.1	+.1	-.15
11W5	0	0	+.45	-.7	+.1	-.15	+.55	-.9	+.1	-.15
15W4	-.1	-.05	+.35	-.55	+.05	-.1	+.55	-.8	+.1	-.15

Record from 9.3.63 : 15.35 hrs.
Mean wind direction S (180°)
Range 90° to 240°

mean wind speed	23 m.p.h.	$q = 1.4$ lb/ft^2
frequent gusts to	39 m.p.h.	$q = 3.9$ lb/ft^2
max. gusts to	45 m.p.h.	$q = 5.0$ lb/ft^2

Gauge	Mean wind		Normal gusts				Extreme gusts			
	press. lb/ft^2	coeff.	press. range lb/ft^2		coeff.		press. range lb/ft^2		coeff.	
15E3	-.45	-.3	+.1	-.9	0	-.25	+.1	-1.8	0	-.35
11E1	-.1	-.05	+.1	-.8	0	-.2	+.1	-1.0	0	-.2
11E2	-.35	-.25	0	-.9	0	-.25	+.1	-1.35	0	-.25
11E3	-.2	-.15	0	-.55	0	-.15	+.2	-1.0	+.05	-.2
11E4	-.2	-.15	+.2	-.8	+.05	-.2	+.45	-.9	+.1	-.2
11E5	-.1	-.05	+.2	-.8	+.05	-.2	+.55	-1.25	+.1	-.25
11W1	-.55	-.4	+.35	-2.0	+.1	-.5	+2.1	-2.7	+.4	-.55
11W2	0	0	+1.1	-1.5	+.3	-.4	+1.9	-1.9	+.4	-.4
11W3	0	0	+.8	-.9	+.2	-.25	+1.6	-1.0	+.3	-.2
11W4	0	0	+.7	-.45	+.2	-.1	+1.25	-.55	+.25	-.1
11W5	0	0	+.45	-.45	+.1	-.1	+.7	-.55	+.15	-.1
15W4	0	0	+.2	-.2	+.05	-.05	+.35	-.55	+.05	-.1

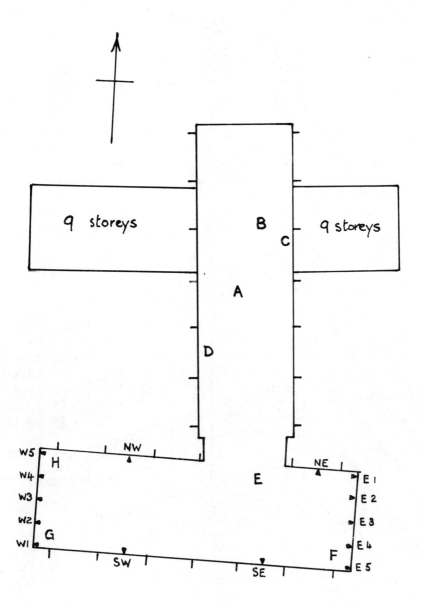

Fig.1. Plan of State House showing gauge positions.

Fig. 2. Part of State House from the south-west.

Fig. 3. Wind-pressure gauge in 15th floor window of State House.

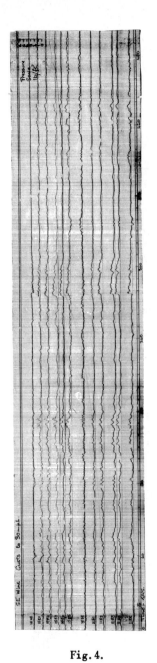

Fig. 4.

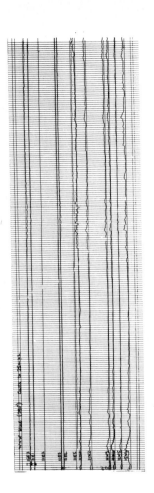

Fig. 5.

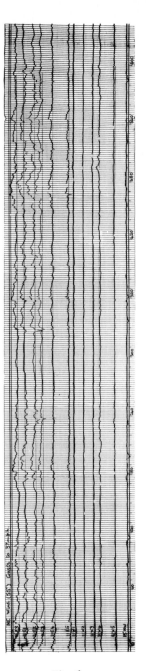

Fig. 6.

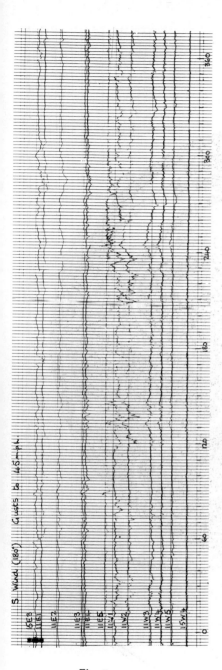
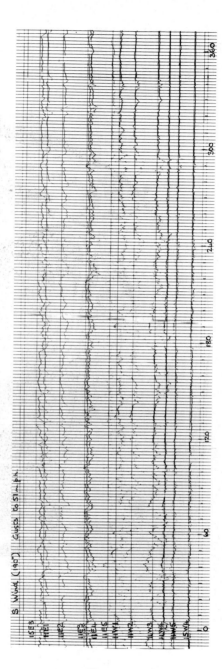

Fig. 7a. Fig. 7b.

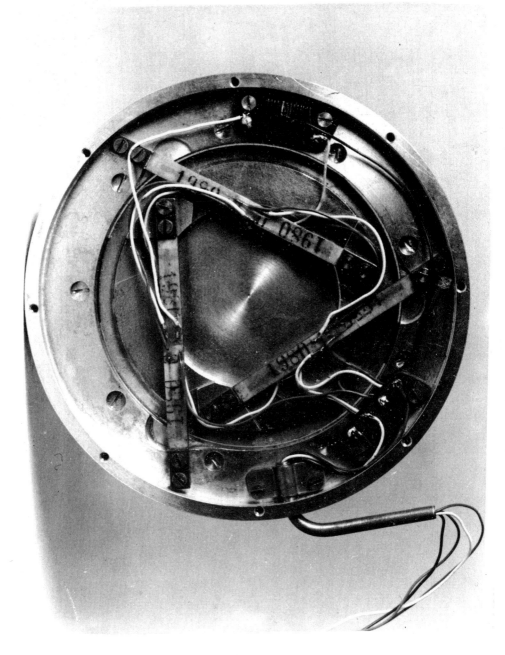

Fig.8. Rear view of wind-pressure gauge with back cover removed.

DISCUSSION ON PAPER 13

PROFESSOR PAGE mentioned a theory which had been developed for ventilation problems which enabled the internal pressure of a building to be calculated. This theory uses an electric circuit analogy. He suggested that the theory could be developed to determine the build up of internal pressure in gusting winds.

MR. ROWE mentioned the similarity between gusting winds and air blasts and some experiments he had carried out with a shock wave passing over elementary building forms. With regard to the problem of a reference pressure in measurements on buildings he thought that the best solution was to take absolute pressures and relate them at the same time to the common reservoir pressure and to the pressure inside the room itself.

MR. RIMMER enquired whether the distortion of the window frame under wind would alter the accuracy of the gauge.

MR. NEWBERRY (in reply) doubted whether the electrical analogy mentioned by Professor Page was sufficient for the purpose of the measurements he had undertaken. He thought also that the analogy with blasts suggested by Mr. Rowe might be dangerous because of the evident fundamental differences between the two flow régimes. In reply to Mr. Rimmer he said that the gauges had been designed to keep acceleration effects to a minimum and neither acceleration effects nor distortion of the frame interfered with the accuracy of the gauge in its present use.

PAPER 4

DETERMINATION DE L'ACTION D'UN VENT TURBULENT SUR LES BATIMENTS ET CONSTRUCTIONS

by

RAYMOND PRIS
Ingénieur E.C.P.
Docteur - Ingénieur
(Paris, France)

DETERMINATION DE L'ACTION D'UN VENT TURBULENT SUR LES BATIMENTS ET CONSTRUCTIONS

by

RAYMOND PRIS

(Paris, France)

AVANT PROPOS

DE nombreuses recherches ont été effectuées en plusieurs pays en vue de déterminer l'importance et la valeur des caractéristiques des courants aériens:

- Vitesse moyenne pour un intervalle de temps donné.

- Evolution de cette vitesse en fonction de la hauteur au dessus du sol.

- Facteur de rafales, rapport des vitesses maximum à la vitesse moyenne.

- Evolution de ce rapport avec la hauteur.

- Etendue des rafales, en rapport avec les dimensions des bâtiments.

- Fréquence des rafales.

- Nature de la turbulence au sol.

- Evolution de la turbulence avec la hauteur.

Ces études ont pour base un nombre variable de relevés météorologiques, expressions d'une situation locale et momentanée; les extrapolations dans l'espace et dans le temps ont néanmoins un grand intérêt; il est etabli que les caractéristiques précédentes relèvent de lois de nature statistique pouvant être utilisées dès maintenant par les constructeurs et que des recherches ultérieures rendront plus précises. Un programme établi d'un commun accord par les services intéressés devrait porter sur les points suivants:

- localisation des stations météorologiques,

- dispositifs d'enregistrement,

- tracé des diagrammes,

- utilisation des diagrammes; détermination des caractéristiques locales,

- extrapolation des résultats.

L'étude suivante a pour but de préciser les résultats obtenus à l'heure actuelle et de présenter certaines suggestions concernant ces diverses questions.

1. NATURE DES COURANTS AERIENS

Classification

Une confusion est souvent faite dans le choix des expressions utilisées pour caractériser la nature d'un courant aérien dont les effets sur les bâtiments et constructions dépendent de l'évolution conjointe plus ou moins rapide des vitesses et des directions. Ces dernières semblent avoir une moindre influence que celles dues à la vitesse; elles n'ont pas encore fait l'objet d'études suivies.

La terminologie suivante est proposée:

    1 - Coups de vent et rafales.

    2 - Vent tourbillonnaire.

    3 - Courants et vents turbulents.

    4 - Vent laminaire.

1 - Coups de vent et rafales

(a) Diagrammes d'enregistrement.

Les volumes interessés par les coups de vent et les rafales sont importants, comparés à ceux des bâtiments. Elles donnent naissance à des vagues aériennes de grande amplitude, englobant l'ensemble de la grande majorité des édifices.

Les vitesses que leur correspondent, sont inscrites en particulier sur les diagrammes établis par les stations météorologiques françaises, *fig.1 et 2*. Les cylindres des enregistreurs effectuent un tour en 24 heures et le déplacement des diagrammes est de 1/4,6 mm par minute, soit 1/276 mm par sec. On relève *fig.1*, entre 10 h et 16 h et pour la hauteur de 39 m, une évolution des vitesses comprises entre 12 et 23 noeuds, soit 6,2 et 11 m/sec; une valeur moyenne peut être définie sur le tracé ainsi qu'une valeur majorée ou valeur d'utilisation, plus rapprochée des valeurs de pointe. Elles s'appliquent à un intervalle de temps plus ou moins long évalué en secondes, en minutes ou en heures, car, du fait de la faible vitesse de déroulement du cylindre, le diagramme ne peut donner aucune indication sur les caractéristiques instantanées du courant. On relève, par exemple, 8,6 m/sec comme vitesse moyenne sur 1 mm du trace de la *fig.1* et, pendant l'intervalle de temps correspondant, les molécules d'air ont parcouru 8,6 x 276 = 2380 m, longueur très supérieure à tout effet turbulent.

L'examen de l'enregistrement de la *fig.2*, anémomètre à 150 m de hauteur, conduit aux mêmes conclusions; l'amplitude des rafales est analogue.

(b) Diagrammes des vitesses moyennes et d'utilisation.

Ils sont établis par extrapolation en partant des diagrammes d'enregistrement, après établissement de formules algébriques de formes exponentielles ou logarithmiques, en fonction de la hauteur au dessus du sol. Leur diversité est bien l'expression des difficultés éprouvées pour établir une règle générale à ce sujet.

Le gradient correspond à l'angle de la tangente à cette courbe avec la verticale, soit $\Delta V/\Delta Z$. Maximum au sol, sa valeur est sensiblement nulle pour des hauteurs dépassant 500 m. La *fig.3* représente de tels diagrammes et *fig.4* en admettant une vitesse de 20 m/sec à la hauteur $Z = 100$ m; elles résultent de l'application d'une formule algébrique (Règles françaises Neige Vent) *fig.3* ou logarithmique (relevés danois, *fig.4*)[1][2].

2 - Vent tourbillonnaire

Le courant correspondant à cette formation est parfaitement défini en mécanique des fluides; toutes les molécules décrivent des cercles autour d'un axe ou d'une ligne ondulée, *fig.5*. Au centre, il se forme sous l'effet de la viscosité un noyau à l'intérieur duquel les vitesses augmentent avec le rayon r, $V/r = C^{te}$; en dehors du noyau et en fluide peu visqueux tel que l'air, l'évolution des vitesses est inversée, et a pour expression $V \times r = C^{te}$. La pression est minimale et les vitesses sont maximales en bordure du noyau.

La représentation naturelle d'un tourbillon est une trombe dont l'axe se rapproche de la verticale, les vitesses maximum peuvent atteindre exceptionnellement 150 à 200 m/sec a la périphérie du noyau, ce qui corespond à une dépression de 2,5 m de hauteur d'eau en application de l'égalité de Bernouilli.

3 - Courants et Vents turbulents

(a) Etude en soufflerie - Couche limite à la surface des corps
Formation de la turbulence.

Les six photos de la *fig.6* obtenues au laboratoire aérodynamique de Göttingen [3], font partie d'un film établi en utilisant un appareil se déplaçant à la vitesse locale du fluide. Elles montrent la naissance de la turbulence et la formation de la couche limite dans un courant au contact d'une plaque plane unie.

Vue 1 - Une première perturbation apparait à un moment donné, au contact d'une irrégularité ou pour tout autre cause.

Vue 2 - Les vitesses diminuant du fait de la viscosité et du frottement pour s'annuler au contact de la plaque, *fig.3 et 4*, la perturbation s'enroule dans le sens des aiguilles d'une montre sur la figure. Un premier tourbillon d'axe horizontal est formé; son intensité et sa longueur dépendent de valeurs aléatoires que l'on ne peut déterminer. La vitesse moyenne u est réduite par rapport à la vitesse initiale V, *fig.7*.

Vues 3 et 4 - De nombreux tourbillons continuent à se former; leur intensité augmente, ainsi que l'épaisseur de la couche.

Vues 5 et 6 - Des aigrettes apparaissent au contact de cette couche et du fluide supérieur non tourbillonnaire. Elles proviennent de la dissipation des tourbillons aux dépens de l'énergie cinétique de ce fluide. Un relevé des vitesses moyennes dans la couche limite est donné *fig.8* sur des plaques de rugosités différentes, à 80 cm du bord avant et pour une vitesse de soufflerie de 20 m/sec.

(b) Etude en vraie grandeur - Vent turbulent.

(1)- La turbulence à la surface d'une plaque et la turbulence du vent sont liées à la notion de viscosité. Si l'air était un fluide parfait sans frottement, l'effet du sol serait réduit et les variations de vitesse auraient pour origine des différences de pression ou de température agissant sur des volumes étendus; les courants thermiques sont toutefois peu sensibles au sol pour les vents forts qui intéressent le constructeur. Les couches inférieures de l'atmosphère parcourent des centaines, sinon des milliers de kilomètres, en partant d'une zône de haute pression; du fait de la viscosité, il y aura création de tourbillons à axes horizontaux dont l'intensité augmentera en approchant du sol comme il a été montré *fig.6* pour la formation d'une couche limite et la vitesse moyenne sera diminuée; mais la différence fondamentale existant entre couche limite et vent naturel réside dans la valeur du gradient au contact de la plaque ou du sol. La répartition des vitesses dans les couches inférieures de l'atmosphère est influencée par la longueur du chemin parcouru et le gradient au sol $\Delta V/\Delta Z$ est très inférieur à celui mesuré pour une couche limite, *fig.9*; les échanges d'énergie sont plus faibles de sorte que le théorème de Bernouilli peut être appliqué pratiquement jusqu'au sol, conduisant à la formule donnant la valeur de la pression dynamique:

$$p = \frac{\rho V^2}{2} = \frac{V^2}{16} \; m^2/sec \; .$$

(2)- <u>Intensité de la turbulence naturelle</u>. Cette turbulence relativement fine ne peut être enregistrée que sur des cylindres à déroulement rapide. Le déroulement du diagramme *fig.10* correspond à 1 mm/sec, soit 276 fois ceux réalisés *fig.1 et 2*. Ces diagrammes ne sont pas comparables.

Le diagramme de la *fig.10* relevé à 10 m de hauteur montre que même en plaine, les courants aériens sont formés au sol par une succession d'ondes multiples, d'intensité et d'amplitudes différentes. Une courbe moyenne peut toujours être tracée, dont les ondulations viendraient recouvrir les courbes de la vitesse du vent présentées dans les Règles.

(3)- <u>Turbulence à proximité immédiate du sol</u>. La turbulence et les écarts de vitesse sont plus accentués à l'approche du sol; on peut même observer exceptionellement une inversion dans leur évolution comme le montrent les enregistrements de la *fig.11*, correspondant à deux anémomètres placés l'un AP à 10 m de hauteur, l'autre AS à 4 m. On lit aux mêmes instants [4]:

$$V \; m/sec \; \begin{cases} AP - 16 \quad 14 \quad 3 \quad 10 \quad 5 \quad 0{,}5 \; - \; (Z = 10 \; m) \\ AS - 19 \quad 18 \quad 8 \quad 15 \quad 12 \quad 10 \; - \; (Z = 4 \; m) \end{cases}$$

(4)- <u>Turbulence en hauteur</u>. Au dessus de 10 m, la turbulence relativement fine, dont il vient d'être question, diminue d'intensité plus rapidement que les rafales. Une étude du Professeur Sherlock [5] donne une cartographie des vitesses enregistrées par des anémomètres à faible inertie montés le long d'un mat de 80 m de hauteur; les vitesses sont comprises entre 30 et 50 pieds/sec (13 a 22 m/sec) et au déroulement rapide du cylindre correspond quatre lectures par seconde. Un simple coup d'oeil sur cette carte permet de juger de la complexité de la répartition des vitesses et explique l'énoncé de conclusions parfois imprécises. Les courbes de vitesse déduites de cette carte pour des hauteurs fixes de 36 - 86 - 136 et 186 pieds (11 - 26,5 - 40 et 57 m) sont reproduites *fig.12 et 13*. Une turbulence très fine est nettement enregistrée à 11 m, pouvant correspondre à des longueurs d'onde de l'ordre de 1 m. Elle semble disparaitre rapidement avec la hauteur; il subsiste ensuite des ondes de plus grande amplitude ou des rafales.

4 - Vent laminaire

Au dessus d'une certaine hauteur, l'écoulement de l'air n'est plus influencé par l'action du sol, sauf dénivellations importantes. En l'absence d'effets thermiques, les couches d'air s'écoulent à une vitesse égale suivant des surfaces parallèles et sans mélange, ce qui est la caractéristique des écoulements laminaires. La preuve en a été apporteé en montant une sphère au bout d'une tige à l'avant d'un avion et en mesurant la résistance pour différentes vitesses. Le changement de régime caractéristique de l'intensité de la turbulence se produit pour un valeur du nombre de Reynolds correspondant à un régime laminaire.

2. ENREGISTREMENT DES PRESSIONS DYNAMIQUES
TURBULENCE ET RAFALES

1 - Principe des mesures

1° - Lecture des diagrammes

Les bâtiments et constructions doivent être stables pour les plus grandes vitesses du vent, en excluant toutefois celles dues à des tornades ou trombes très violentes mais de très faible fréquence. On estime que, du fait de l'inertie des ensembles, les tourbillons ou pointes de rafales de grande intensité mais de faible durée peuvent être supportée sans dommage; il n'y aurait pas lieu de tenir compte des pointes des tracés sur les diagrammes des figures 1 et 2, mais il n'est pas certain que l'on aboutisse à des conclusions identiques après étude du diagramme à déroulement rapide de la *fig.10*.

2° - Vitesses et pressions

Le constructeur est intéressé non par la valeur de la vitesse du vent mais par les pressions exerceés sur les parois, proportionnellment au carré des vitesses.

3° - L'influence des variations rapides de la direction
n'a pas encore été étudiée. Il y aurait lieu d'établier un dispositif

enregistreur ayant les caractéristiques suivantes:

- instantanéité de l'enregistrement;
- enregistrement des pressions dynamiques;
- enregistrement des directions du vent dans un plan vertical et dans un plan horizontal.

2 - Appareillage

Les conditions précédentes amènent au choix d'une sphère comme récepteur, reliée à des dynamomètres munis de dispositifs de mesure électrostatiques *fig.14*.

    F   étant la résistance,

    c_t le coefficient de la sphère,

    S   la surface de référence.

On a pour la pression dynamique q :

$$q = \frac{F}{c_t \cdot S}$$

Les lectures ne sont possibles que si le coefficient a une valeur constante indépendante de la vitesse, ce qui ne peut être réalisé avec les surfaces polies. Contrairement à tous les essais antérieurs, la sphère utilisée devra posséder une forte rugosité ou une nervuration amenant une réduction importante du nombre de Reynolds critique. Cette influence est bien marquée par la *fig.15* pour des cylindres [6][7]. Alors que pour une surface polie (courbe 11), le coefficient c_t est encore évolutif pour un produit $V \times D$ égale a 30 m²/sec, il résulte d'essais éxécutés au N.P.L. sur cylindres recouverts de toile émeri (courbes 4-5-6) que cette constance peut être obtenue à partir de $V \times D = 4$ [8].

Les vents forts, de vitesse au moins égale à 15 m/sec devant être seuls enregistrés, on aurait pour diamètre de la sphère, en admettant un écoulement analogue:

$$V \times D = 4 \qquad D = \frac{4}{15} = 0,27 \text{ m} .$$

La vitesse de déroulement du diagramme devra être de 1 mm/sec au minimum.

3 - Montage - *fig.16*

Une station comportera un certain nombre de ces enregistreurs de pression permettant d'établir une carte des vitesses analogue à celle représentée *fig.12*. Tenant compte de la direction des vents dominants lesquels sont souvent les vents les plus forts, des mats supportant les enregistreurs seront disposés dans un plan perpendiculaire à cette direction. Ces enregistrements permettront de déterminer l'étendue et la fréquence des tourbillons et des rafales, les directions locales, ainsi que l'influence de la hauteur au dessus du sol.

3. ACTION DE LA TURBULENCE ET DES RAFALES SUR LES BATIMENTS EN VRAIE GRANDEUR - MESURE DES PRESSIONS SUR LES PAROIS

1 - Turbulence, rafales et bâtiments

(a) <u>Définitions</u> - Il n'existe pas de limite définie séparant turbulence et rafales.

Pour un bâtiment cubique dont une face est normale au vent, *fig.17*, on peut dire qu'une fluctuation est turbulente lorsqu'elle n'intéresse qu'une partie du bâtiment, alors qu'une rafale l'englobe entièrement suivant les trois directions OX - OY - OZ.

Dans le premier cas, les pressions devraient être rapportées à des vitesses de référence différentes pour obtenir des coefficients comparables; dans le deuxième cas, la vitesse unique est celle de la rafale.

(b) <u>Cas de l'onde turbulente</u> - L'onde désignée par la lettre A, *fig.10*, est reportée dans le diagramme vitesses - chemin parcouru ou longueur d'onde, *fig.18*. On voit que, à un moment donné, sur des longueurs égales à 40, 25, 10 m, les vitesses réelles auront pour valeur 8, 9, 10 m/sec.

Action sur l'ensemble d'un bâtiment:-

On pourrait dire que à cet instant, la vitesse moyenne du vent sera un peu supérieure à 10 m/sec pour l'ensemble d'un bâtiment cubique de 10 m de coté, et de 9,30 m/sec pour un bâtiment cubique de 40 m de coté. De plus grands écarts seraient certainement observés en d'autres points du diagramme.

Action sur une façade:-

En admettant une égale étendue de l'onde suivant les axes OZ et OY, *fig.17*, le raisonnement ci-dessus serait encore applicable en fonction de la longueur de la façade.

Déformation de l'onde:-

En réalité, les conclusions précédentes ne sont pas entièrement justifiées par suite de la déformation de l'onde au passage du batiment, effet plus marqué avec les faibles longueurs d'onde. Les éléments au vent, les façades en particulier, seront plus influencés que les grands ensembles; ce serait le contraire pour les éléments élevés d'un bâtiment de grande hauteur.

(c) Cas de la rafale - Le professeur Sherlock [5] donne pour longueur d'onde 8 fois la profondeur du bâtiment pour être en droit de considérer une vitesse unique pour tous les éléments d'un bâtiment.

2 - Mesure des pressions sur les parois

(a) <u>Conditions générales</u> - La détermination des pressions présente certaines difficultés ayant pour causes:

l'évolution de la vitesse et de la direction du vent, en particulier en fonction de la hauteur;

une forte turbulence à l'approche du sol;

l'existence d'une direction déterminée correspondant aux vents forts, lesquels permettent un enregistrement précis;

l'obligation de connaître avec une grande précision la pression barométrique (ou pression statique) permettant le calcul ou la mesure des pressions relatives à la surface des parois;

Exemple:- Une étude très étendue a été entreprise aux Etats-Unis pendant plusieurs années avant 1940, comportant la mesure des pressions aux 36 ème, 56 ème, et 76 ème étages de l'Empire State Building [9]. Elles étaient relevées avec des manomètres à liquide, par rapport à une pression intérieure aux mêmes étages. Les auteurs signalent les difficultés éprouvées pour déterminer les vitesses de reférence applicables aux différentes prises.

(b) Organisation des essais - En conclusion d'une recherche récente [4], deux méthodes d'essais peuvent être proposées:

(i) Enregistrement des efforts résultants.

Les prises de pressions montées sur les parois enregistrent la différence des pressions existantes sur les faces extérieures et intérieures. La vitesse du vent est relevée par la station météorologique la plus proche, suivant les règles habituelles. Les relevés de pressions et de vitesses effectués en des moments correspondants sont inscrits sur des tableaux portant sur une longue durée, ainsi que sur un grand nombre de bâtiments de formes diverses situés dans une même région. La lecture des tableaux fera connaître la valeur des pressions maximum susceptibles de se produire en fonction de la vitesse météorologique, laquelle, ici, ne constitue qu'un repère. Toutes mesures de gradient et de turbulence sont supprimées.

(ii) Détermination des coefficients de pression extérieurs et intérieurs.

Les pressions relatives devant être relevées, il est nécessaire de prévoir un dispositif établissant une pression de référence fixe. La preparation des mesures comporte les éléments suivants: *fig.19*:

> un volume étanche à pression constante mesurée par un baromètre à eau,
>
> des prises de pression extérieures et intérieures,

et pour l'étude du vent, des appareils enregistreurs:

> un anémometre mesurant la pression dynamique, *fig.14*, ainsi que sa direction,
>
> une prise statique (ou barométrique).

Ces appareils restent à créer. Les prises de pression seront mises en liaison avec le volume étanche. Les enregistrements seront instantanés. Tous les éléments nécessaires au calcul des coefficients sont enregistrés.

Remarques

Les enregistrements seraient facilités en montant les prises à une hauteur suffisante au dessus du sol pour que les effets de fortes turbulences soient évités; seules les rafales auraient une action.

Les mesures sur constructions cylindriques ont un intérêt particulier. La pression sur la génératrice au vent donne la valeur de la pression dynamique ainsi que la direction du vent, *fig.20*. Toutefois, l'état de surface influe considérablement sur la valeur des pressions, *fig.21, 22 et 23*.

4. CONCLUSIONS GENERALES

1 - Nature des courants aériens

Couche limite sur plaque plane et gradient des vitesses du vent ont une même cause initiale, la viscosité de l'air; mais les effets sont très différents en raison de la longueur du chemin parcouru par les courants aériens. Le gradient au sol est très inférieur a celui observé sur une plaque unie.

Le frottement amène la formation de tourbillons à axes horizontaux près du sol et de rafales sur une plus grande hauteur, avec perte d'énergie cinétique variant en tout point de l'espace. La carte des vitesses est d'une grande complèxité et relève de valeurs aléatoires. Le courant est turbulent près du sol.

2 - Enregistrement des pressions dynamiques - Turbulence et rafales

Un appareil adapté à l'étude aérodynamique des bâtiments doit enregistrer la pression dynamique du vent, ainsi que les directions. Ces enregistrements doivent être instantanés. L'emploi d'une sphère à surface rugueuse ou nervurée liée à des dynamomètres éléctrostatiques parait indiqué; ces sphères seraient montées à différentes hauteurs dans un plan normal aux vents forts les plus fréquents.

3 - Action de la turbulence et des rafales sur les bâtiments.
Mesure des pressions sur parois en vraie grandeur.

On peut utiliser soit une méthode simplfiée comportant uniquement la mesure des pressions résultantes sur les parois; les valeurs maximum seraient mises en regard des relevés météorologiques régionaux. Une méthode complexe est nécessaire pour la détermination des coefficients de pression; un appareillage serait à créer.

BIBLIOGRAPHIE

1. Règles définissant les effets de la neige et du vent sur les constructions - Paris, 1946.

2. JENSEN, M. - The Model Law for phenomena in natural wind - Copenhague, Ingénioren - International edition, vol.2, 1958.

3. PRANDTL, L. - Guide à travers la mécanique des fluides - Trad. Monod, - Paris, Dunod, 1952.

4. Centre Expérimental de recherches et d'études du Bâtiment et des Travaux publics. Mesure de l'effet du vent sur les bâtiments de grande envergure - Paris, 1962. (dossier 415).

5. SHERLOCK, H. - Variation of wind velocity and gusts with height, dans Proc. Amer. Soc. civ. Engrs., 1952, 126.

6. PRIS, R. - Influence de rugosité et de la nervuration sur la résistance aérodynamique des cylindres, dans Annales de l'Institut technique du bâtiment et des travaux publics, nov. 1960.

7. PRIS, R. - Résistance de cylindres à base circulaire et à surface rugueuse, Ibid., juillet-aout 1961.

8. National Physical Laboratory. Influence de la turbulence et de la rugosité sur la résistance d'un cylindre. Dans Report et Memoranda, No.1283, 1929-1930.

9. RATHBURN, Ch. - Wind Forces on a tall Building. Trans. Amer. Soc. civ. Engrs., 1940, 2056.

10. Report of the Building Research Board. 1961. Londres, H.M. Stat. Off. 1962.

11. PRIS, R. - Action du vent sur les bâtiments et constructions. Préparation et montage des maquettes en soufflerie. Dans Annales de l'Institut technique du bâtiment et des travaux publics, fevrier 1962.

12. BARTH, R. - Mesure au tunnel aérodynamique de la résistance au vent d'éléments cylindriques jumelés, dans der Stahlbau, juin 1960.

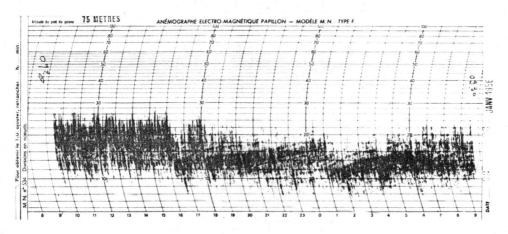

Fig.1. Hauteur: 39 m - Déroulement: 24 heures

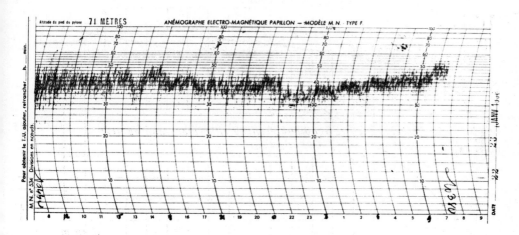

Fig.2. Hauteur: 150m - Déroulement: 24 heures

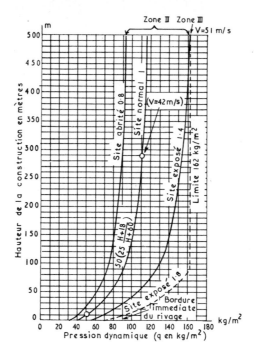

Fig.3. Variation de la pression dynamique ($q = \dfrac{V^2}{15}$) en hauteur et suivant la zone et le site

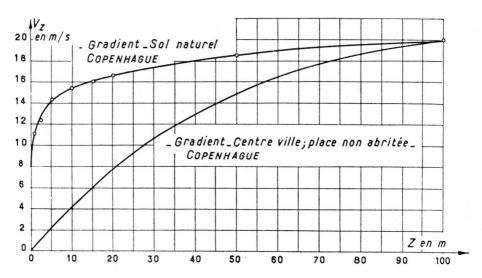

Fig.4. Gradient des vitesses du vent relevé au-dessus d'un sol cultivé, sans dénivellations importantes. On a admis une vitesse égale à 20 m/s à la hauteur z = 100 m.

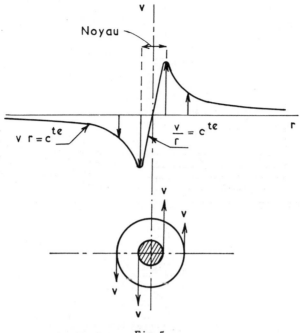

Fig. 5.

Fig. 6. Propagation d'une perturbation turbulente

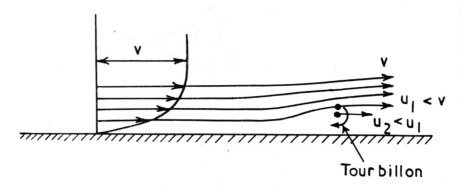

Fig. 7.

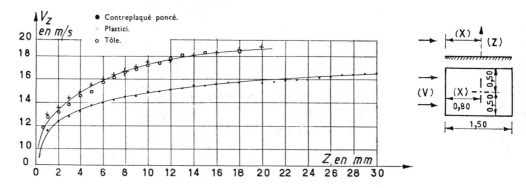

Fig. 8. Relevé des vitesses à l'intérieur de la couche limite formée au contact de différentes surfaces. Son épaisseur caractérise la rugosité de la surface. Après un parcours de 80 cm, son épaisseur minimale est de 20 mm pour une surface lisse (tôle, plastici).

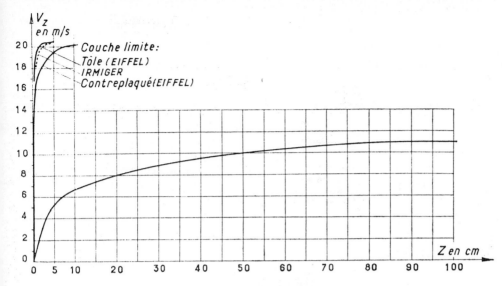

Fig.9. Comparaison des gradients, vent naturel et courant de couche limite sur plaque-sol. Les évolutions sont totalement différentes. Le gradient naturel est beaucoup moins accentué que celui obtenu sur plaque-sol (hauteurs z en centimètres).

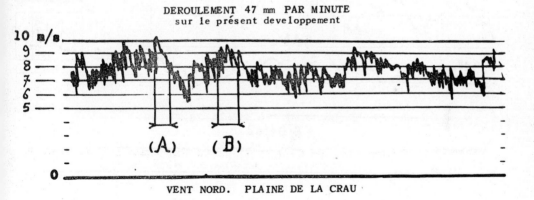

Fig.10

Diagramme 7

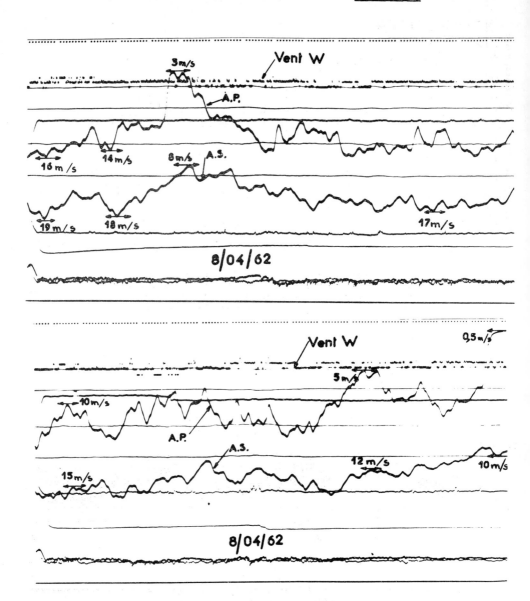

Fig.11. Variations exceptionnelles d'évolution de la vitesse
du vent à 10 m (A.P.) et à 4 m (A.S.) (vent W)

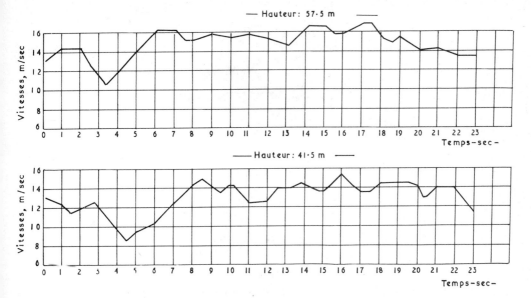

Fig.12

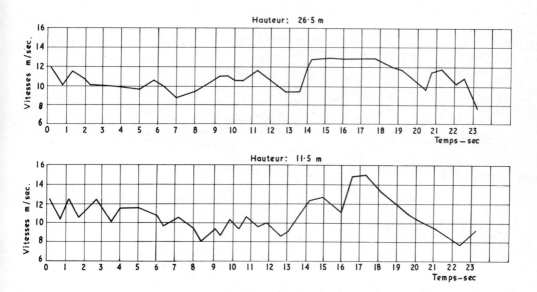

Fig.13. Cartographie des vitesses

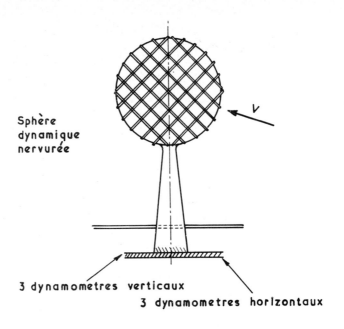

Fig.14

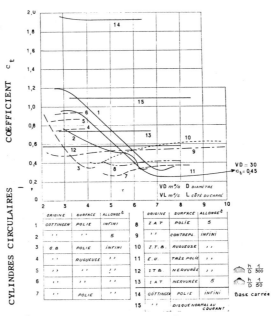

Fig.15. Étude de la similitude. Évolution du coefficient de résistance c_t. Paramètres influant sur la similitude: produit VD m²/s, turbulence du courant.

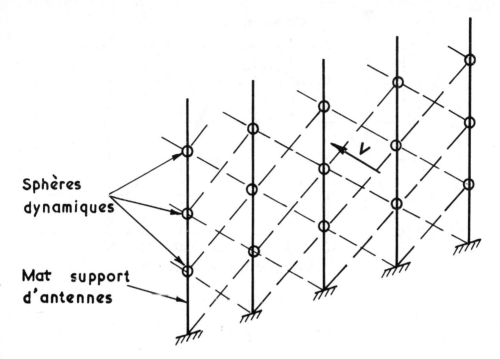

Fig.16

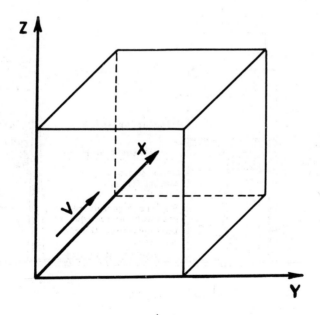

Fig.17

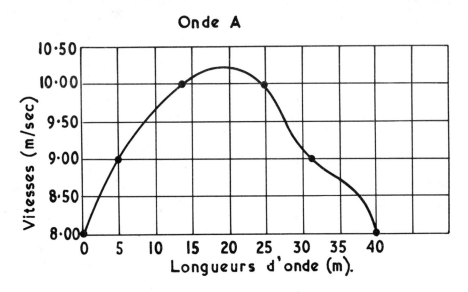

Fig.18

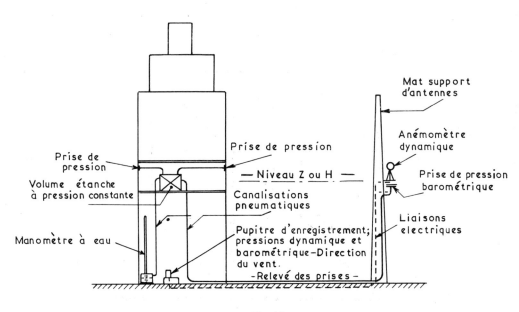

Fig.19

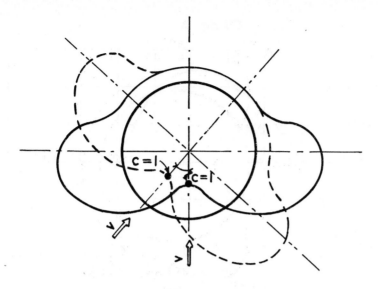

Fig. 20

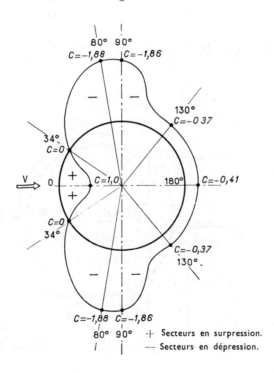

Fig. 21. Cylindre en élancement infini. Diagramme des pressions. Surface = tôle peinte - c_t = 0,420.

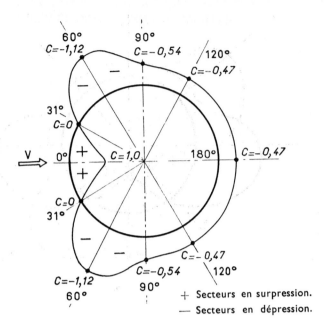

Fig.22. Cylindre en élancement infini. Diagramme des pressions. Surface = rouille profonde - $c_t = 0,622$.

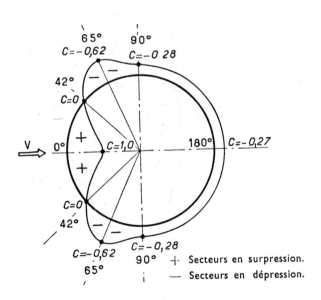

Fig.23. Cylindre en élancement infini. Diagramme des pressions. Surface = ciment très rugueux - $c_t = 0,612$.

CHAIRMAN'S CONCLUDING REMARKS

SIR GRAHAM regretted that no time was left for discussion of Dr. Pris' paper.

It was strange that meteorologists have always measured wind pressures and then laboriously converted them into velocities, although their application was mainly as pressures. Beaufort, in devising his Beaufort wind scale was not concerned with velocities, but about pressures exerted, e.g. "which no canvas can withstand".

The meteorologists had been accused of paying very little attention to the wind scale required by structural engineers. The reason for this is that the great deal of knowledge we now have about wind structure near the ground was not obtained because of the necessity of studying its effect on buildings, but from the necessity to study the diffusion of gases and smoke in the atmosphere. The interest was therefore in low wind speeds and not in the high speeds of interest to the structural engineer.

WRITTEN DISCUSSION
PAPER 4

MR. MORRISON. I am surprised the Dr. Pris is not familiar with the design of the ERA Gust/Anemometer described in ERA report C/M06 and which formed the subject of a paper presented to the I.E.E. by H. H. Rosenbroch in 1951.

The sensing element used in connection with the instrument was a sphere, perforated to give a roughness which was shown to be necessary during wind tunnel tests at N.P.L. The instrument had a response, to a change in wind-speed representing full scale, of 0.1 seconds and a damping ratio of 0.64 of critical.

Recently two other gust anemometers have been developed by the ERA which have the same response and damping characteristics, but where the emphasis on design has been on weight and cost per instrument. The spheres used in these anemometers are perforated table tennis balls and the amount of damping required to give the correct ratio was found to be extremely large. One of these anemometers is an omni-directional instrument designed to give an output linear against wind pressure, whilst the other gives a cosinusoidal output about either of two preferred directions $100°$ apart and again the output is linear against wind pressure.

In view of the trouble we have experienced in obtaining the correct damping ratio for these instruments we would be extremely interested to learn how Dr. Pris has obtained the necessary damping ratio and response times in connection with a sphere having a diameter of 25 cms.

SESSION 3

Chairman:

Mr. E. N. Underwood
(President, Institution of Structural
Engineers, U.K.)

PAPER 15

MODEL LAW AND EXPERIMENTAL TECHNIQUE FOR DETERMINATION OF WIND LOADS ON BUILDINGS

by

NIELS FRANCK
(Copenhagen, Denmark)

MODEL LAW AND EXPERIMENTAL TECHNIQUE FOR DETERMINATION OF WIND LOADS ON BUILDINGS

by

NIELS FRANCK

(Copenhagen, Denmark)

1. INTRODUCTION

THIS paper deals with a description of an experimental technique used at the Wind-Laboratory, the Technical University of Denmark, at model tests for determination of wind loads on buildings.

The experimental technique described has permitted to produce turbulent boundary layers in the wind tunnel. With respect to turbulence, the model tests could therefore be brought into agreement with the conditions in nature which is a condition of obtaining accordance between model tests and nature.

2. THE MODEL LAW

The condition of establishing similarity of turbulence in the wind tunnel and in nature is that of the relation between the roughness of the surfaces in the wind tunnel and in nature.

The wind in the layer near the ground is nearly always turbulent the whole way down to the surface.

The turbulence in the flow alone depends on the roughness of the surface when the atmosphere is in a neutral equilibrium, i.e. when there are no thermic contributions to the turbulence.

The velocity profile in a turbulent flow over a rough surface follows the logarithmic formula

$$\frac{v(z)}{v_*} = \frac{1}{\mathcal{K}} \log_e \frac{z + z_o}{z_o}$$

v(z) is the velocity at the height z,
v* is the friction velocity,
$\mathcal{K} = 0.4$ is Kármán's constant, and
z_o is the roughness parameter.

The roughness parameter z_o is a measure of the roughness of the surface concerned and, at the same time, of the turbulence in the boundary layer over the surface so that a more turbulent boundary layer corresponds to a surface of greater roughness.

Fig. 1 shows a velocity profile measured in nature. The abscissa is the relative velocity, the velocity at a height of 100 m being fixed at 1.0. The ordinate is the height above the level of the ground marked out on a logarithmic scale.

In this way the velocity profile becomes a straight line. If the line is extended to intersect the ordinate axis, the value of z_o is obtained. The profile in *fig. 1* was measured over a cultivated field, and the roughness parameter is $z_o = 0.95$ cm.

The condition of obtaining similarity between the phenomena in the model tests and the corresponding phenomena in nature is that the roughness parameter for the tunnel bottom, z_o, must be to scale to the roughness parameter in nature, Z_o.

$$\frac{z_o}{Z_o} = \frac{d}{D}$$

where d and D are corresponding dimensions of the object in the tunnel, respectively of that in nature.

Obviously, the model must be entirely immersed in the turbulent boundary layer of the tunnel bottom.

The model law for phenomena in natural wind has been practised at the Wind Laboratory since 1948, where its validity has been demonstrated by experiments for phenomena depending upon the wind velocity, such as shelter effect, as well as for phenomena depending upon the velocity pressure.

The model law was first published by Martin Jensen: Shelter Effect, Copenhagen, 1954.

The satisfaction of the model law necessitates a number of special requirements of the laboratory equipment used.

In the following, an explanation will be given regarding the principles to be observed for the construction of apparatus and experimental technique at model experiments with houses in a turbulent boundary layer.

3. APPARATUS

3.1 Wind Tunnel

The experimental section of the wind tunnel must be closed, and of such length that it is possible to produce a turbulent boundary layer of sufficient thickness over the tunnel bottom.

At the Wind Laboratory tests were made in a wind tunnel of the type with enclosed working section and open return flow. The tunnel which is shown in *fig. 2* is 60 x 60 cm in cross section and its total length is 10 m. The insertion of a 2 m long extra section between sections 1 and 2 permits, however, the length of the tunnel to be increased to 12 m, which was done at the experiments described in this paper.

The experimental section of the tunnel consists of a framework with hatches which together with the frame form the inside of the tunnel. It is provided with honeycombs at each end.

Maximum wind velocity in the tunnel is 32 m/s.

3.2 Tunnel Coatings.

In model tests, to reproduce the situation in nature, it must be possible to establish velocity profiles in the wind tunnel with such roughness parameters as are to scale to those in nature. This has been practised by various coatings of the tunnel bottom.

Below, five tunnel coatings are described as have been applied at our model tests of wind loads on houses.

In all cases the coating extended from the honeycomb at the inlet end of the tunnel to the 4th experimental section (see *fig. 2*). The coating covered the entire width of the tunnel.

The thickness of the turbulent boundary layer corresponding to the tunnel coating evidently increases with the distance from the inlet end of the tunnel.

The thickness, t, of the boundary layer in the 4th section of the tunnel is noted for each of the five tunnel coatings.

Velocity profiles over the tunnel bottom measured in the 4th section have been plotted for each coating on *fig. 3*.

The abscissa is the relative velocity, the velocity at 100 cm level being fixed at 1.0. The ordinate is the logarithm of the height above the surface.

The different tunnel coatings are shown in *figs. 4, 5, 6, 7* and *8*.

Smooth masonite, *fig. 4*. 0.3 cm thick sheets of hard masonite were screwed to the bottom of the tunnel. $z_0 = 1.8 \times 10^{-3}$ cm, t = 12 cm.

Sandpaper, *fig. 5*, was glued to the tunnel bottom. $z_0 = 1.85 \times 10^{-3}$ cm, t = 15 cm.

Corrugated paper, *fig. 6*, was fixed to the tunnel bottom. $z_0 = 0.047$ cm, t = 18 cm.

Broken stones, *fig. 7*. Cubical stones of sizes between 1.5 and 2 cm were distributed over the tunnel bottom. z_o = 0.21 cm, t = 20 cm.

2.5 x 2 cm fillets. *fig. 8*. Wooden fillets, 2.5 cm in height and 2 cm in width, were fastened to the bottom. The fillets were placed transversely to the tunnel, forming angles varying between 90° and 70° to the longitudinal axis of the tunnel. The mutual distance was varied fortuitously between 15 cm and 25 cm. The velocity profile in *fig. 3* corresponds to z_o = 0.6 cm, the thickness of the boundary layer in the 4th section is t = 28 cm.

3.3 Pitot-static Tube

The velocity profile of the wind in the tunnel is measured by means of pitot-static tubes and manometer.

Measurements of the velocity profiles in the different boundary layers in the tunnel necessitate the use of a pitot-static tube of such size as to permit measuring of velocities near the surface of the tunnel coating without disturbance of the movements of the air within the area of measurement.

At the Wind Laboratory the pitot-static tube shown in *fig. 9* was used for measurements of velocity profiles.

The two branches are tubes of external diameter 0.1 cm. The short branch is open at the end so that the pressure will there be the sum of the dynamic and static pressures.

The larger branch protrudes 0.3 cm in front of the short branch and is closed at the end but provided with two lateral holes, d = 0.009 cm, abreast of the point of the short branch. From these holes the static pressure is taken.

The velocity pressure will thus be registered by a manometer, the two sides of which are connected with the two branches of the pitot-static tube.

The pitot-static tube was mounted in the manner shown in *fig. 10* allowing it to be moved in the vertical symmetry plane of the wind tunnel and adjusted with an accuracy of 0.01 cm.

3.4 Models of Houses

The size of the models placed in the tunnel must be chosen so that the narrowing of the cross-section of the tunnel, due to the placing of the model, will not cause any systematic error in the measurement.

The models of the houses are placed at the bottom of the tunnel, which, in fact, substitute the surface of the ground, and they must be completely immersed into the turbulent boundary layer.

The models used at the Wind Laboratory were made of brass sheets with smooth and plane surfaces. All models were hollow.

The roofs and walls of the models were provided with measuring holes at such a number as to permit the obtaining of a picture in detail of the distribution of the wind load.

The points of measurement in the roofs of the models were located in the greatest number near the corners where experience shows that the largest suction values and the greatest gradients occur.

At the points of measurements thin steel tubes, bore 0.1 cm, were placed normal to the surface of the model. *Fig. 11* shows a section of one of the house models used and a detail of a point of measurement.

To the free end of the steel tube a rubber hose was attached to transmit the pressure to a manometer.

3.5 Turntable

The models of the houses placed in the wind tunnel must be revolved in a convenient way in relation to the direction of the wind.

For this purpose we used a turntable which was placed on the same level as the bottom of the tunnel in the 4th section of the wind tunnel. The turntable has a diameter of 21 cm, and on a graduated scale engraved along the edge, the angle between the direction of the wind and the model could be measured.

The rubber hoses were led from the points of measurement through a tube at the centre of the disk and out of the tunnel to the multimanometer described below.

3.6 Multimanometer

For the sake of survey it is most expedient that the conditions of pressure at several of the holes of measurement of the model could be registered simultaneously. In this way the variation of the load on the surfaces of the model could more easily be surveyed. Hence, a multimanometer is a necessity.

At the Wind Laboratory a multimanometer with 15 tubes was used. The manometer liquid was alcohol of specific gravity 0.80, and the manometer tubes were mounted with an inclination corresponding to a factor of 0.4 of the hydraulic head in mm.

4. EXPERIMENTAL TECHNIQUE

The experimental technique applied at our model tests for determination of the wind load on a model of a house is described in detail in this section.

(a) The turntable is mounted at the bottom of the tunnel in the 4th section and a suitable coating is placed across the whole bottom.

(b) Under the ceiling of the 3rd section of the wind tunnel and above the turbulent boundary layer a Prandtl pitot-static tube is mounted, with which, in connection with an ordinary FUESS manometer, the velocity of the air flow above the boundary layer is measured.

(c) The velocity profile in the 4th section is determined by means of the pitot-static tube described in section 3.3.

Furthermore, the usually small difference between the static pressure in the turbulent boundary layer at the location of the model, and that at the location of the Prandtl pitot-static tube above the boundary layer is determined.

(d) The model is fastened to the turntable, and 15 points of measurement, distributed for instance over a wall, are connected with the upper ends of the manometer tubes by means of rubber hoses.

(e) The static pressure of the Prandtl pitot-static tube is connected with the top side of the reservoir of the multimanometer.

(f) Preceding measurements proper, a survey of the variations of pressure and suction at the 15 points of measurement is made by observation of the manometer while the model is turned in relation to the direction of the wind.

In this way it is possible to find the points of particular interest with regard to wind load, and also the distribution of the wind load could be determined in the main.

(g) The load at each of the points of measurement is determined as the mean of the loads at different wind velocities.

The loads (pressure or suction) found by the tests are made dimensionless by division with a specific velocity pressure. At all our measurings we have used the velocity pressure which occurs in the turbulent boundary layer at the level of the ridge.

This results in the shape factors for each of the points of measurement.

The velocity pressure at the level of the roof of the model is arrived at as the velocity pressure at the Prandtl pitot-static tube in the 3rd section multiplied with a factor that is determined as the ratio between the velocity pressure at the level of the roof without model in the tunnel and that at the Prandtl pitot-static tube, measured simultaneously.

The computation of the c-values must be corrected for the difference that may occur between the static pressure at the Prandtl pitot-static tube and the static pressure at the model, since it is really the latter that should be applied to the top side of the reservoir of the multimanometer.

Obviously, it would have facilitated the measurings if the pitot-static tube of reference could simply be placed in the actual boundary layer at the level of the roof of the model. The tunnel used at the Wind Laboratory is, however, too narrow for such placing of the pitot-static tube of reference to permit entirely correct results.

5. EXAMPLE OF DETERMINATION OF WIND LOAD

In this section, as an example, test results for a model of a house with desk roof will be brought, comprising investigations in boundary layers with small as well as great turbulence.

The length of the model is 140 mm, the width 70 mm, and the height to the lower eave 70 mm. The inclination of the roof is 1 ÷ 10.

In *figs*. 12, 13, 14 and 15 measuring results of wind loads on the top surface of the roof of this model are given. 45 points of measurement were spread over the area of the roof.

The results shown in *figs*. 12 and 13 are measured for the model placed in a boundary layer with small turbulence, smooth masonite sheets at the tunnel bottom.

The results shown in *figs*. 14 and 15 apply to the model placed in a boundary layer with great turbulence, 2.5 x 2 cm fillets at the bottom of the tunnel.

The shape factors given in the figures are $c = \frac{p}{q} \times 100$, where p is pressure or suction at the point of measurement concerned, and q is the velocity pressure in the turbulent boundary layer, measured at the level of the upper line of the roof.

Curves have been traced through points indicating equal c-values, end extreme point values are stated. Areas signifying suction corresponding to c greater than 40 are hatched with variated shades.

In the figures the shape factors are shown for the positions of the model which are the most interesting in respect of wind loads.

For certain positions of the model, the load on a part of the roof only is given, viz the part for which the load assumes extreme values while for the rest of the roof the load is trivial.

The very great suction values which occur at the corners of the roof for the wind approaching the corner at an angle of about 45° to the walls should be noted. Particularly large suction values will occur when the high corner of the roof is directed towards the wind. Suction values corresponding to c = 420 in small turbulence were measured here.

Also, it should be noted that there are considerable differences between the c-values in the two different turbulent boundary layers, which may serve as an illustration of the significance of the turbulence with regard to the wind loads.

For wind directed towards the high edge of the roof at 45° to the wall it will be noticed that the greatest suction will occur when the model is located in small turbulence.

For wind directions lengthwise or crosswise to the model, the greatest suction will, however occur at the windward edge of the roof when the model is placed in great turbulence.

From this it will appear that maximum loads may occur for the model located in small as well as in great turbulence.

(A complete report of wind load tests will be published at the end of 1963, entitled Martin Jensen and Niels Franck: Model Tests in Turbulent Wind, Part II).

REFERENCES

JENSEN, Martin, Shelter Effect, Copenhagen, 1954.
JENSEN, Martin, The Model Law for Phenomena in Natural Wind. - *Ingeniøren, International Edition*, 1958, 2, 4.
JENSEN, Martin and FRANCK, Niels, Model Tests in Turbulent Wind, Part I, Copenhagen, 1963 and Part II (under preparation).

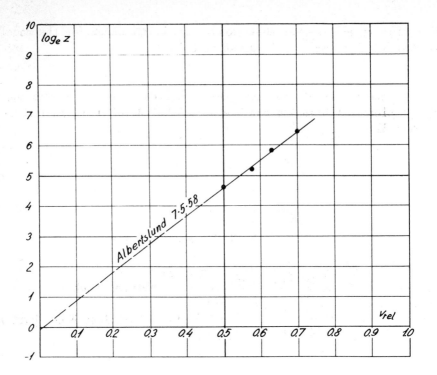

Fig.1. Velocity profile measured in nature.

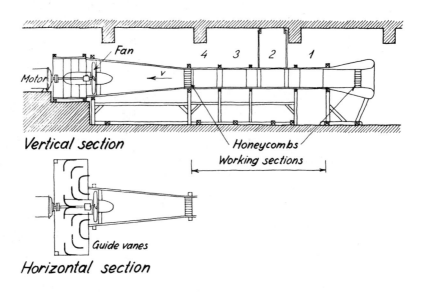

Fig.2. 60 x 60 cm wind tunnel.

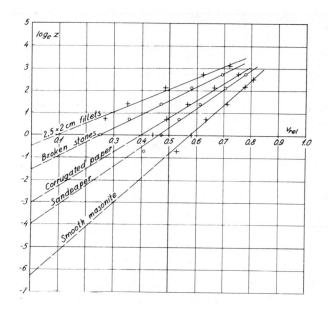

Fig.3. Velocity profiles over five tunnel coatings.

Fig.4. Smooth masonite.

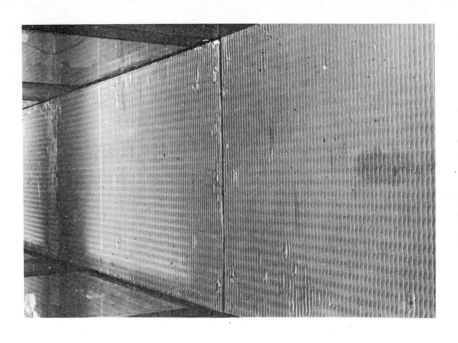

Fig. 6. Corrugated paper.

Fig. 5. Sandpaper.

Fig.8. 2.5 x 2 cm fillets.

Fig.7. Broken stones.

Fig.9. Pitot-static tube for precision measurements. Length of the longer horizontal branch is 1.2 cm.

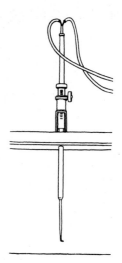

Fig.10. Holder for the pitot-static tube in Fig.9. By means of the holder the pitot-static tube may be placed at any point of the vertical symmetrical plane of the tunnel.

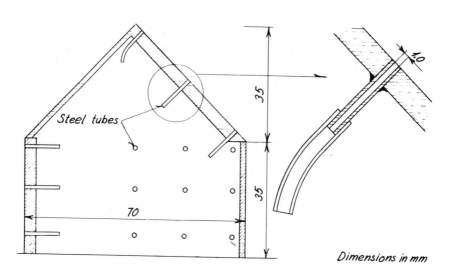

Fig.11. Model of a house used in the experiments. Detail of point of measuring.

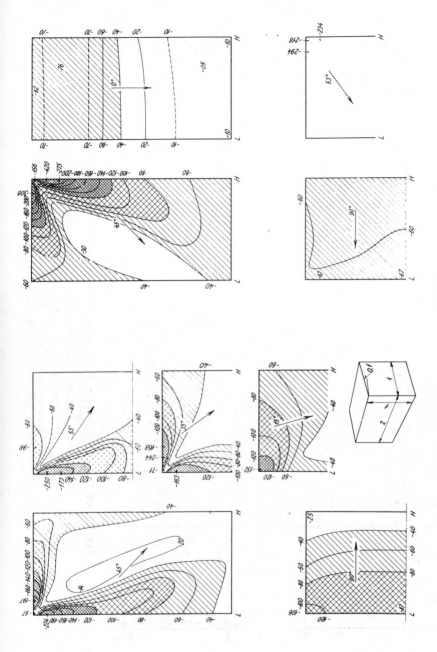

Figs. 12 and 13. Test results for a model of a house with desk roof (inclination 1 ÷ 10).
Shape factors c are given for the top surface of the roof.
$c = \frac{p}{q} \times 100$, where p is the suction and q is the velocity pressure at the level of the upper line of the roof.
Figs. 12 and 13 apply to the model placed in small turbulence.

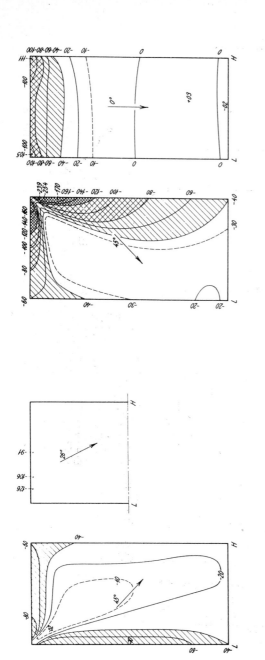
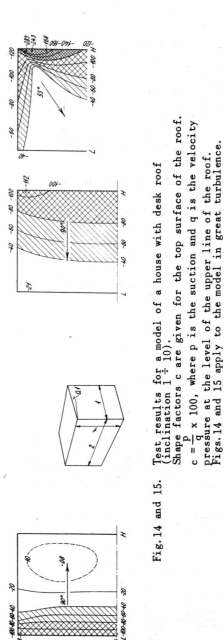

Fig. 14 and 15. Test results for a model of a house with desk roof (inclination 1 ÷ 10). Shape factors c are given for the top surface of the roof. $c = \frac{p}{q} \times 100$, where p is the suction and q is the velocity pressure at the level of the upper line of the roof. Figs. 14 and 15 apply to the model in great turbulence.

PAPER 6

EFFECTS OF VELOCITY DISTRIBUTION ON WIND LOADS AND FLOW PATTERNS ON BUILDINGS

by

W. DOUGLAS BAINES
(University of Toronto, Canada)

EFFECTS OF VELOCITY DISTRIBUTION ON WIND LOADS AND FLOW PATTERNS ON BUILDINGS

by

W. DOUGLAS BAINES

(Department of Mechanical Engineering, University of Toronto, Canada)

SUMMARY

PRESSURE distributions on models of walls and rectangular-block structures have been measured in a wind tunnel. The tests were conducted both in an artificially produced velocity gradient, used to simulate natural conditions, and in a constant velocity field, for comparison with standard procedures. Several rules have been deduced from the results by which pressure distributions and wind loads can be predicted for buildings in any specified wind field. Changes in the flow pattern due to the velocity distribution have been observed and correlated with the pressure distribution. Only on the front surfaces were distinct phenomena observed: an intense downward flow carrying high energy air to ground level exists on tall buildings, and on low structures a region of reversed flow occurs at ground level resulting in an attached eddy.

1. INTRODUCTION

In many instances of contemporary design practice, wind loads are assumed to be the result of a steady wind with constant velocity at all elevations above ground level. Under these assumptions the pressure p at any point on the structure can be related to the ambient pressure p_o and the constant velocity V_o through the pressure coefficient

$c_p = \dfrac{p - p_o}{V_o^2/2}$ where ρ is the ambient density of the air.

If the structure has sharp corners and wind velocities are of the order of 5 mi./hr. or more, the theory of flow around bluff bodies indicates that c_p is a unique function of the wind direction relative to the building. For a few simple shapes the values of c_p can be deduced from certain classical experiments in aerodynamics. However, for the large majority of practical cases, wind-tunnel tests are required for each indiviual structure shape and orientation with respect to the wind direction. Thus there results a catalogue of pressure coefficients for every building geometry. While there is no all-inclusive listing of these coefficients, a very large number of building shapes have, for instance, been tested at the Iowa Institute of Hydraulic Research, and the results correlated and summarized[1]. Furthermore, average values for building surfaces have also been assembled from various sources and issued in handbook form for design purposes[2]. However, it is widely recognized that any extraordinary shape must still be studied in scale model tests in a wind tunnel.

The commonly adopted testing procedure neglects two major meteorological characteristics, i.e. the gustiness of the wind and the variation of velocity with height. These phenomena are not independent but are the natural consequence of the friction of the earth's surface on the motion of large air masses. It can easily be deduced that the rougher the surface the more gusty (unsteady) the wind and the greater the curvature in the velocity distribution.

Turbulence of the natural wind field has been measured in the field and found to have the same general properties as wind-tunnel turbulence. However, there is a very large difference in the size of the eddies. In a conventional wind tunnel the eddies are an order of magnitude or more smaller than the size of the model, whereas in the field the eddies are an order of magnitude larger than the structure. The effect of these large eddies on wind loads has been discussed by Davenport[3] who has shown that the dynamic properties of the building influence the resulting wind loads. A few simple cases have been analyzed but much remains to be done. It is evident from Davenport's work that conventional wind-tunnel tests do not give an indication of the size of these dynamic loads.

The natural velocity distribution can, however, be readily reproduced in a wind tunnel and the loads resulting from this effect analyzed. Screens and grids can be installed in the tunnel test section and any velocity distribution produced. Pressure distributions can be readily measured and local values of the pressure coefficient c_p obtained. The reference velocity V_o used here is that of the undisturbed wind field at the elevation of the top of the building. The inclusion of still another variable (i.e., the velocity distribution) on which the pressure coefficients are dependent further complicates the presentation of test results. A given plot will now show results for only one shape, one orientation and

one distribution so that many tests of a particular building would be
required to give a complete picture. There are, however, trends observed
in tests on simple shapes which can be of real help to designers in estimating the loads on buildings of more complex shape.

This paper presents the most significant results of a programme of
measurements of pressure distribution on typical building shapes in a low-
speed wind tunnel. Two wind fields were used. One of constant velocity
represented typical standard wind-tunnel test conditions; for the other a
velocity distribution approximating that in the atmospheric boundary layer
was developed. The distribution selected was typical of urban conditions as
defined by Davenport[3]. The models in every case had sharp corners and
rectangular plan section to minimize scale effects. Attempts were also made
to apply the principles of flow around bluff bodies to give analytical
procedures for the prediction of pressure distribution for any velocity
field. However, this approach was successful only for tall buildings. For
low buildings, i.e., with heights roughly equal to the lateral dimensions
or less, the pressure distribution cannot be predicted, but total loads
can be predicted to the accuracy common in practice.

2. PRESSURE DISTRIBUTION AROUND BUILDING SHAPES

Virtually all building shapes are classified as bluff bodies and thus
have distinctly different pressure distributions from streamlined bodies
such as air foils and submarine hulls. The distinguishing property is the
wake of separated flow surrounding the rear part of the body. As sketched
on *fig.1* the flow approaching the body is deflected by the front facing
surfaces but separates completely from the surface at a sharp edge. This
is a direct consequence of the inability of the fluid to undergo the very
large accelerations required to follow the surface at the corners. Because
of shear along the edge of the wake a return flow toward the body in the
centre of the wake is induced. Thus, the characteristics of the wake
region are (i) velocities much smaller than the mean flow, and, hence,
almost uniform pressure on the body surfaces, (ii) a gentle flow upstream
along the surfaces and (iii) a negative pressure compared to the ambient
since it is dictated by the prevailing pressure at the point of separation.
In addition to this mean flow there are intense pressure and velocity
fluctuations resulting from the entrainment process along the wake streamline. It should be noted that these fluctuations are so large that the
position of this streamline can be defined only by a long time average.
Beyond this general description little can be said about the absolute
values of c_p for a given body shape. In general, the location of separation
depends on the shape and size of the body, the orientation and velocity of

the wind but for bodies with sharp edges, i.e., for most buildings, the separation line is fixed along the edges. Hence c_p varies only with wind orientation for any given shape.

The surfaces of a structure exposed to the mean flow display a pressure distribution identical to that on a body in inviscid fluid flow with a free streamline for the wake streamline. Thus the pressure is above ambient over most of the frontal surface and there are steep pressure gradients near the edges where streamline curvature is sharp. For irrotational flows, e.g., in a constant velocity field, the pressure distribution can be predicted from theory but in many cases this requires a long, detailed calculation. For the boundary-layer velocity distribution, theory has not yet been developed, hence experimental measurements must supply the missing information.

There will of course be a thin boundary layer on these front surfaces. For the Reynolds numbers of model and full-scale buildings the thickness is of the order of 0.1 of the width of the building. The presence of this boundary layer does not affect the pressure distribution. However, since the velocity within the layer is small, it is influenced by the pressure distribution. Thus a secondary flow results along the surface of the body in the direction of the maximum pressure gradient. This fact can be used to predict the direction of the surface flow from the contours of c_p. Lines drawn normal to the contours thus are directions of secondary flow. Farther away from the body the streamlines are not in the direction of the pressure gradients because of the higher inertia of the main flow.

This discussion intimates the wide applications of the pressure contours such as are presented in *figs.2 and 3*. Some of these are

(i) evaluation of wind loads,
(ii) illustration of the direction of flow of pollutants released from the building,
(iii) determination of direction of surface motion of deposited rain,
(iv) general shape of snow drifts around building. This requires measurement of pressure distribution on the ground surrounding the building.

Such a list shows the practical advantages to be gained from any summarization of pressure distributions on building shapes.

3. EFFECTS OF A NATURAL VELOCITY DISTRIBUTION ON FLOW PATTERN

As a consequence of the preceding discussion it is to be expected that the potential flow and wake regions will be influenced in very different ways by changes in the velocity distribution in the approaching flow. Separation lines will obviously not be affected but the shapes of wake streamlines will be changed because of the change in the potential flow field. Thus the only change of significance to be expected on the body surfaces in the wake will be in the average level of c_p, because differences in the pattern of pressure contours will be small. On the other hand, the front faces will directly reflect the variation of stagnation pressure with height above the ground. Indeed, experimental observations reveal two flow phenomena not found in a constant velocity field. These are

(i) a secondary flow down the front of the body due to the fact that the pressure near the top is larger than at the bottom,

(ii) a reversal of flow along the ground in front of the body. This is the result of the adverse pressure gradient built up ahead of the body acting on the slower moving fluid near the ground. The result is a flow separation of a different kind on the ground at a point ahead of the body which involves a lateral flow away from the separation point to both sides. The size of flow reversal zone relative to the height of the building varies with the steepness of the velocity gradient, i.e., the ratio of stagnation pressures at the top and bottom and also with the height-width ratio of the building. The latter is an important parameter because the length of the zone of adverse pressure gradient is roughly equal to the width of the building. This upstream flow combines with the main flow on the front of the structure to produce the eddy sketched on *figs. 7 and 9*.

The explanation of these phenomena by theoretical principles is straightforward, and similar reasoning can also be used to predict the effects on the pressure distribution. The secondary flows alone have little effect but the flow reversal and the consequent eddy give a zone of relatively constant pressure similar to the wake. This effect changes the pressure distribution over the entire front face. In a manner similar to flow in the wake, the size of the flow reversal zone fluctuates with time since the point of separation is not fixed by geometry. Thus large pressure and velocity fluctuations are found over the front surfaces.

The foregoing description of the potential flow zone was deduced from careful observation of a wall placed perpendicular to the flow. Examination of other wind orientations on the wall and the front faces of other shapes showed similar effects except that the size of the eddy reduced as the angle of orientation was decreased. In the interests of brevity this paper will be restricted to the 90° orientation, data and descriptions of other orientations being contained in other publications[4] [5].

In the following paragraphs the measured pressure distribution on some typical shapes will be discussed in detail and numerical values assigned to the effects described above. These can then be used to predict the pressure distributions on other shapes and provide rules for framing specifications.

4. WIND-TUNNEL TESTS OF PARTICULAR SHAPES

Description of Tests

Pressure measurements were made on models constructed of acrylic plastic sheet material in the low-speed wind tunnel of the Department of Mechanical Engineering, University of Toronto. This is an open-return tunnel with a 4 ft. by 8 ft. cross-section and a maximum speed of 25 ft./sec. Thus, although the pressures to be measured were small and required the careful use of micromanometers the results did not require correction for wall effects. The largest dimension on any model was 18 inches, and consequently any blockage effect was negligible. The methods and instrumentation used in the tests were conventional in wind-tunnel practice except in the creation of the velocity distribution. In every case a detailed measurement of the pressure distribution was made with the model in a constant velocity field. For this the model was set on a thin ground board mounted clear of the natural boundary layer of the tunnel floor. These results were taken as standard and agreed well with published data.

The boundary-layer velocity distribution was created in the lower half of the tunnel by installing a curved screen in the entrance of the test section. Considerable effort was expended in shaping this screen to produce the desired distribution. The final shape was roughly parabolic with the vertex upstream at the tunnel centre line. The top half of the cross-section was filled with a continuation of the screen normal to the flow direction. The action of a screen in producing a shear flow has been explained and a theoretical derivation presented by Elder[6]. However, in the present instance it was not possible to calculate the resulting flow profile because of the interaction with the natural tunnel boundary layer and the large velocity deviation from the average. The velocity distribution evolved was described by the polynomial law used by Davenport[2].

$$\frac{V}{V_o} = \left(\frac{y}{y_o}\right)^{1/k} \tag{1}$$

where V_o is the velocity at the reference elevation y_o. Most of the tests were conducted with a distribution for which $k = 4$. This corresponds to conditions on the outskirts of a large city. For a few of the earlier tests a different screen was used which produced a distribution for which $k = 6$. It was found that constant handling of this screen produced creases and wrinkles which acted like small lenses. Flow concentration resulted which could not be corrected and thus the screen had to be rebuilt.

Tall Buildings

Very clear evidence has been obtained[4] of the effects on a building of square floor plan with a height-width ratio of 8. This is typical of contemporary skyscrapers. All of the changes in pressure distribution between the two velocity distribution studies can be readily explained by elementary theory. These effects are best illustrated in *figs.2 and 3* which present the pressure contours for the wind normal to the front face. For this orientation the front face is in the potential flow region and all other surfaces are in the wake.

In the constant velocity field the pressure distribution on the front, see *fig.2*, is typical of that on a very long flat plate. The contours are vertical except near the top where over a length of about 2B (B = width) an end effect is displayed. In this area some of the mean flow deflected by the building passes over the end and some over the sides.

On the lower part of the building the flow deflection is entirely lateral and at the top the deflection is entirely vertical. In *fig.4* the streamlines close to the centre line are schematically presented. These were deduced from the pressure distribution and checked with tufts of wool held in the air flow. On the sides, roof and back of the structure the pressures are more uniform but still show gradients much larger than those found on low buildings[1]. Examination of the pressures in detail shows that the roof has more suction than the sides and back near the base. This indicates that the flow diversion over the top has resulted in more sharply curved wake streamlines. Since the wake is relatively stagnant an upward flow is induced from the relatively high pressure zone near the ground. The normals to the contours indicate that this takes place up the sides and back and into the roof zone from where the fluid is removed by the intense shear layer.

The patterns of pressure contours and flow lines are changed considerably in the presence of the velocity gradient, note *fig.3*. There is a strong downward flow and pressure gradient on the front, the pressures on the roof and sides are quite uniform and the roof pressure is about half the intensity of the constant velocity case. These can all be explained in terms of the variation of stagnation pressure with height on the front of the structure. In the region immediately upstream of any bluff body the flow must decelerate in the longitudinal direction and accelerate in the lateral direction for the fluid to pass around the body. Along the central plane this lateral acceleration must be small for reasons of symmetry, hence an area is developed on the front of the body where the pressure approaches the stagnation pressure

$$p_s = \rho \frac{V_o^2}{2} + p_o \quad \text{i.e.,} \quad c_p = 1.00.$$

In the constant velocity case this is an area extending about 6B of the height from the base. The same tendency exists in the flow with a velocity distribution but the stagnation pressure varies with height. Thus the layer of air close to the surface is subjected to a strong vertical pressure gradient and a strong flow down the structure results as shown on *fig.4* for the streamlines near the central plane. As a result of the downward flow the velocity distribution is more uniform immediately behind the building. Thus smaller structures in the immediate vicinity would be subjected to higher wind loads than with the tall structure absent. This is a relatively minor effect, but the higher velocities on neighbouring streets would surely be noticeable, particularly by the movement of debris. Near the top of the building the downward flow is superimposed on the natural upward flow from the end effect. The result is a substantial reduction in the quantity of air flowing over the top and a reduction of the height influenced by the end. Consequently, the suction on the roof is reduced by about half. The effect of the velocity distribution on the pressure distribution near the top is the same as a reduction in height of the building. Thus it would be expected that the relative size of the reduction of suction on the roof would be larger for taller buildings and also for distributions with greater curvature. This is verified in observations on lower buildings. With the roof pressure reduced to the same size as that on the walls and back, the upward flows along these surfaces do not exist.

If it is assumed that a flow with a velocity gradient exists in thin, independent horizontal layers then a simple analytical solution is obtained for the pressure distribution. The pressure at any point is influenced only by the velocity in the main stream at the same elevation. Pressure coefficients are then determined from the equation

$$c_p = c_{p_o} \left(\frac{V}{V_o}\right)^2 \qquad (2)$$

where V = the velocity at the elevation where c_p is determined, and
c_{p_o} = pressure coefficient at the same point on the building in a

constant-velocity field.

This assumption would be valid if the layer on the front of the body were so thin that the main stream were not affected by the downward flows shown on *fig.4*. This layer thickness has not been determined.

The overall accuracy of this linear approximation has been found by plotting profiles of the pressure distribution on the front face as measured and as predicted by Equation (1). *Fig.5* is the vertical profile on the centre line of the front face and this shows clearly the agreement found. Over the lower 75% of the structure, i.e., height equal to six widths, the agreement is very good. There is a small region of constant pressure at the base which is negligible for this case but which is a major factor in the pressures on low buildings. On the top section of height 2B the effect of the reduction in flow over the top is felt. The pressures are about 8% higher than those given by Equation (2). As shown by the vertical pressure contours in this region on *fig.3*, the deviated flow is to the sides. Thus the centre-line pressure excess is closer to $\varrho V^2/2$ than that given by Equation (2). Lateral profiles also show that these deviations from Equation (2) are generally found all across the width. Profiles for $y/H = 0.47$ show good agreement, those for $y/H = 0.735$ show deviations for part of the width and those close to the top show pressures all higher than predicted.

It is instructive to compare the deviations between total load on the front face as predicted by Equation (2) and as measured. The total load is about 3% larger than predicted but the total moment about the base is about 8% larger. Thus it would be expected that for all tall buildings this effect is of real importance.

A simpler approximation indicated from the above discussion is

$$c_p = c_{p1} \left(\frac{V}{V_o}\right)^2 \quad (3)$$

where c_{p1} = a standard pressure coefficient given by the lateral profile for the uniform region in *fig.2*. A single lateral profile is specified and the pressure profile at any elevation is given by multiplying this by the local stagnation pressure. This means that along the centre line for $0°$ orientation the pressure coefficient is $(V/V_o)^2$. From *fig.5* it is seen that this gives good agreement up to within 5% of the top of the building. Similarly good agreement is found on the lateral profiles.

Pressure distributions on a surface in the wake region cannot be expected to be described by Equation (2) or (3). These pressures reflect conditions of separation all along the edges and thus are an integrated effect of the velocity distribution. For the equations it is assumed that pressure is determined by local conditions. An approximation to the wake pressures in a non-uniform velocity distribution can be obtained from the same general reasoning and the measurements in the constant velocity field. The average pressure in this case was influenced by the lowest i.e., the roof, pressure. Thus in a boundary-layer type velocity distribution, if the pressure coefficient at separation were constant, the pressure would be lowest where the velocity was highest. This would make the roof pressure the controlling factor again but as noted above, the roof pressure is affected by the reduction in flow up the building at the top. Consequently the measured reduction of 40% on pressures is consistent with these arguments. Extending this reasoning to the general case leads to the conclusion that average pressure coefficients should decrease with building heights and the relative reduction of c_p by a non-uniform velocity distribution should also decrease. These have been observed.

Low Buildings

The classification of a building as low or high in relation to the pressure distribution on its surface is through the ratio of height to width of the frontal projection. Absolute height is relatively immaterial compared to this ratio in influencing pressure distribution because this is primarily the result of top and side deflection of the main flow streamlines.

(i) Very Wide Wall

This case of a low building demonstrates the opposite limit of effects to the skyscraper. Flow around the sides of the wall is negligible compared to that over the top. Thus for most of the width the flow is essentially two-dimensional and a single curve presents the pressure distribution.

In a constant velocity field the wide wall is subject to the same flow conditions as a long flat plate submerged in an infinite fluid. The plane of symmetry of the plate corresponds to the floor on which the wall rests and the streamlines and pressure distribution shown on *fig.6* are thus obtained. On the front there is a more gradual decrease of pressure from the base to the top than on a tall building, note *fig.5*. On the rear the constant suction of the wake is seen. The value of -1.4 shown for c_p was measured[7] for an infinitely long plate. This changes with height-width ratio, being about -1.0 for a ratio of 0.1.

The flow pattern and pressure distribution in a boundary-layer velocity field shown on *fig.7* have been deduced from measurements on a wall with a height-width ratio of 0.5 and corrections made for end effects. A stable eddy with a horizontal axis forms along the front of the wall at the base. This eddy is roughly circular and the size depends primarily on the curvature of the velocity profile. The more curved the profile, i.e., the smaller the value of k in Eq. (1) the larger the eddy. It forms as a consequence of the adverse pressure gradient ahead of the wall due to stagnation on the wall. A return flow is set up along the floor in response to this pressure gradient because the inertia of the fluid particles near the floor is very small. Because the pressure throughout this eddy is virtually constant the pressure variation over the front of the wall is small and the load on this surface is greater than for a constant velocity field. It appears that the suction in the wake is also not so large because of the same effects which lead to a smaller roof pressure on a tall building in the boundary-layer case. Thus the total load on the wall is reduced in the presence of the boundary layer but by a relatively small percentage. Available results indicate that this reduction is about 10% for a velocity field typical of an urban area. However, because of the uniform pressure on the front the overturning moment about the base of the wall is increased by a small percentage in the presence of the boundary layer.

(ii) Wall of Height-Width 1:1

For walls with the height and width of the same order the flow pattern shows the effects of deflection over both the top and the sides. It is thus a combination of the two simplified cases discussed above; the distribution of pressures varies considerably with height-width ratio. No general conclusions can be drawn applicable to all ratios but trends due to the top and side effects can be seen in the measurements. *Figs. 8 and 9* present the flow pattern and centre-line pressure distributions for the constant velocity and boundary-layer fields respectively. Nothing unusual is seen in the first figure but with a velocity gradient the pattern reflects the phenomena discussed above. The effect noted on the tall building of pressure varying with free-stream stagnation pressure is also found but only over the upper portion. The lower part is covered with the eddy found in front of a very wide wall. The length of this eddy is so short that there is very little recirculation and near the sides the streamlines are entirely lateral. Thus the flow near the wall in this area exhibits the same downwash noted for the tall building. Again the structure acts to bring high velocity air to lower levels. The effect of the eddy is most clearly noted on the floor where a dividing streamline exists. In the layer adjacent to the floor the air downwind of this streamline is moving away from the building and upwind of the streamline toward the building. Thus any material moved along the floor tends to deposit near this streamline in the zone of minimum velocity. A good example of this dividing streamline is the snow drift ahead of a building which accumulates during a high wind. There is always a clear area in the front of the building. From the arguments above it is evident that the size of this area depends on the height-width ratio of the building and the curvature of the velocity profile.

In both velocity fields suction in the wake is much smaller than for the wide wall. There is also little difference between the suction in the two cases. In addition the suction is less than that found on tall buildings, indicating that the wake pressure is least negative for ratios near unity. This is consistent with observations of flat plates in infinite wind streams[7].

(iii) Rectangular-Block Structures

From the preceding discussion it is evident that the effects of velocity distribution on the flow around walls are readily explained and predictions can be made of the pressure distribution on walls of any shape in any wind at any orientation. The only point which remains to be settled in predicting the wind loads on building shapes is the influence of velocity distribution on the side, roof and back surfaces of a building. Because of the separation along front edges the pressure distribution on the front is independent of the depth of the building, i.e., dimension in the flow direction. For wind directions other than the normal incidence, the pressure distribution over the front may vary with length of side walls but the effect of velocity distribution on the pressure distribution is the same as for thin walls.

Measurements of pressure distribution on a cube are typical of all rectangular-block structures. These show the complexity of flow in any velocity field and demonstrate that no simple conclusions about the wake pressures can be drawn. *Figs. 10 and 11* present measured pressure contours for the constant velocity field and a boundary-layer field typical of a suburban area.

Front surfaces exhibit a distribution virtually identical to that measured on a wall of a 1:1 ratio. These are typical of all block structures of similar height:width ratio. On the sides and the roof the suction is slightly greater in the boundary-layer flow and the pattern is quite different. This change is consistent with the differences found in the suction on the rear, i.e. in the boundary-layer flow the suction is much less than in the constant-velocity field. For other structural shapes and wind directions the results were quite similar but some cases have been found in which the differences in pressure distribution between the two velocity fields were very small. The variations are not easily explainable and no doubt are due to the complex nature of the flow in the wake. For example, along the sides the wake is restricted in width by the solid wall. Thus the flow which separated on the front edges may become reattached ahead of the rear edges. This area of reattached flow depends on the velocity distribution as well as the shape and orientation parameters. There is reason to believe that it may also depend on the Reynolds number of the flow, hence the small-scale model tests may not define the pressure distribution on a building. However, the average pressure on the sides and roof is not likely to vary much with a change of Reynolds number of 100 or 1000 fold.

From the results shown on *figs. 10 and 11* and measurements on other rectangular-block structures it appears that the soundest recommendations which can be made at present are maximum suction coefficients applicable to any conditions. For the wind normal to one face these would be a pressure coefficient of -0.55 for the rear surface and -0.70 for the sides and roof.

(iv) Structures with Gable Roofs

Measurements have also been made of the flow pattern and pressure distribution around a cube with a series of gable roofs [5]. In every case the effects of the velocity distribution were identical to those observed on walls and rectangular-block structures. This is not surprising since the roof is located in a region of relatively small velocity gradient. Such a conclusion would not be expected for changes made to the building shape near the ground.

(v) Structures of Intermediate Heights

A progressive change in effects was noted in comparing the pressure distributions on walls for height-width ratio varying from 1 to 8. As the building height increased the size of the frontal eddy remained roughly equal to the width of the building. Thus the pressure distribution above a height equal to the width could be accurately predicted by the schemes used for the tall building. Below this level the pressure is roughly constant. In the wake region the effects of velocity distribution varied from those observed on the cube to those on the tall building.

5. APPLICATION TO BUILDING DESIGN PROBLEMS

Phenomena associated with the gradient of wind velocity have been observed and analyzed from the wind-tunnel studies of small building models. These same effects must occur on full-scale buildings but it would not be expected that the identical pressure distribution would be found. Buildings differ greatly from models in surface finish and appurtenances which particularly influence the line of separation and wake suction. Furthermore the Reynolds number of the flow over the building is up to 500 times that over the model. This means that the areas of flow reattachment will be much different in the two cases. In opposition to these considerations are the facts that forces due to steady winds on any structure need not be predicted within a few percent and the detailed pressure distribution is rarely required. Thus the general rules deduced by comparing the pressure distributions in constant velocity and boundary-layer wind fields should be of sufficient accuracy for the prediction of wind loads in the majority of cases. For buildings of peculiar shape or location either a more careful analysis based on the complete test results[5] or a model test will be required.

It is possible, therefore, to formulate a few simple rules for the prediction of wind loads on a building with flat surfaces in a natural wind field. It is assumed that the pressure distribution determined from a standard wind-tunnel test is known. For the front surfaces:

(i) above an elevation equal to the width of the building the pressure coefficient c_p can be accurately predicted by multiplying c_p for a long flat plate at the same point by $(V/V_o)^2$ where V_o is the undisturbed velocity at the top elevation of the building,

(ii) below this elevation the lateral pressure distribution is relatively constant with height: the distribution is approximately the same as that determined by (i) above at elevation equal to the width,

(iii) if the building is much lower than the width, a constant value of $c_p = 0.9$ provides a conservative loading over the whole front face for a perpendicular wind. At other wind orientations the values listed in building codes suffice.

For the side and rear faces of a building the pressure coefficients given by standard wind-tunnel tests should be used. In some cases it may be possible to reduce the design loads but this rarely exceeds 10%.

For the roof the pressure coefficients can be reduced by up to 50% for a tall building, (height - width ratio = 8) but for a building of ratio = 1 the pressure should be that for a constant velocity field.

Probably the most important conclusion of the studies has been that c_p varies relatively little from the values listed in building codes provided that it is based on the natural velocity at the elevation of the top of the building. This means that the average wind pressure is a function of the height of the building. This is consistent with the observation that a building in a natural wind field acts as a deflector bringing high energy air down to lower levels. Thus neighbouring low structures are subjected to higher velocities and ground level winds increased in magnitude.

In addition to the total loads on walls discussed above, there is considerable practical interest in local high pressures which dictate loads on windows, fasteners and similar small items. A few of these situations have been studied and in no case was a consistent intensification or relief of the condition noticed. Thus, for example, closely spaced buildings of comparable height should be designed for constant velocity conditions.

It is apparent that wind loads on all building shapes are as complex in description as the shapes themselves. This complexity is considerably reduced if the designer is familiar with the characteristics of flow around bluff bodies and can use this knowledge in his deductive process. The prime object of this paper has been the supplying of such knowledge for the case of a natural wind field.

ACKNOWLEDGEMENTS

The assistance of the National Research Council of Canada and the University of Toronto in financial support of the studies described above is gratefully acknowledged. In particular much material has been obtained under a contract with the Division of Building Research, NRC. At all times the staff of the Division have been most helpful in locating problems of interest and discussing the practical implications of the results.

REFERENCES

1. CHIEN, N., FENG, Y., WANG, H.J., and SIAO, T.T., "Wind-Tunnel Studies of Pressure Distribution on Elementary Building Forms", *Iowa Institute of Hydraulic Research, Iowa City, Iowa, U.S.A.*, 1951.
2. SCHRIEVER, W.R. and DALGLIESH, W.A., "Handbook of Pressure Coefficients for Wind Loads", *National Research Council of Canada*, No. 6485, 1961.
3. DAVENPORT, A. G., "The Application of Statistical Concepts to the Wind Loading of Structures", *Proc. Instn. civil Engrs.*, Vol. 19, Aug. 1961.
4. BAINES, W.D., "Effect of Velocity Distribution on Wind Loads on a Tall Building", *University of Toronto, Department of Mechanical Engineering*, TP 6203, June 1962.
5. HAMILTON, G.F., "Effect of Velocity Distribution on Wind Loads on Walls and Low Buildings", *University of Toronto, Department of Mechanical Engineering*, TP 6205, Nov. 1962.
6. ELDER, J.W., "Steady Flow through Non-Uniform Gauzes of Arbitrary Shape", *J.fluid.Mechs.*, 5, 1959, p.355.
7. ROUSE, H., "Fundamental Principles of Flow", Chapter I of "Engineering Hydraulics", *John Wiley and Sons, New York*, 1950.

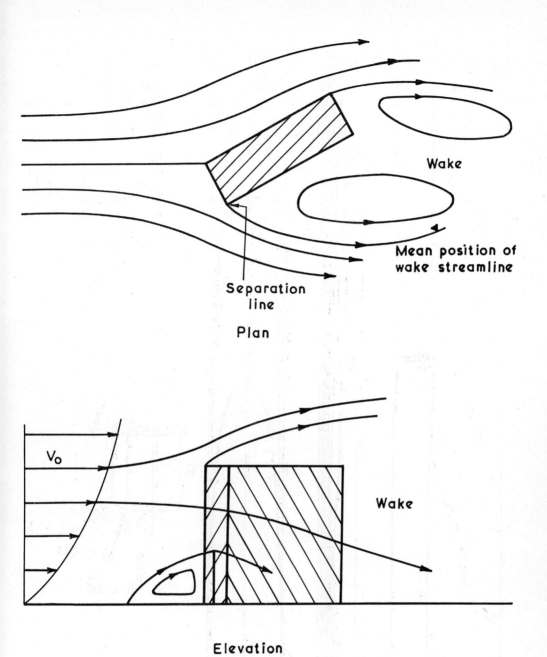

Fig.1. Definition Sketch of Flow over a Building

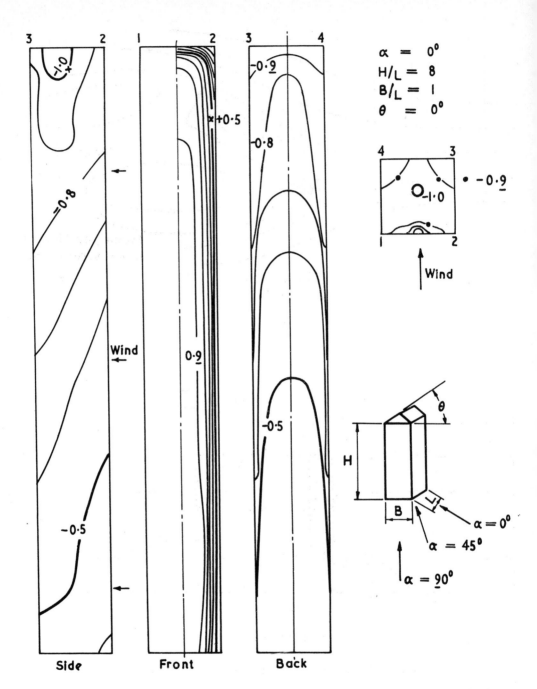

Fig.2. Pressure Distribution, Tall Building in a Constant Velocity Field

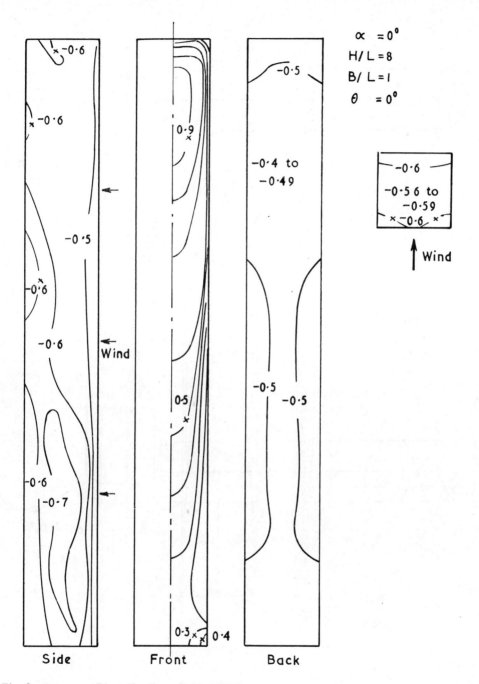

Fig.3. Pressure Distribution, Tall Building in a Boundary-Layer Velocity Field

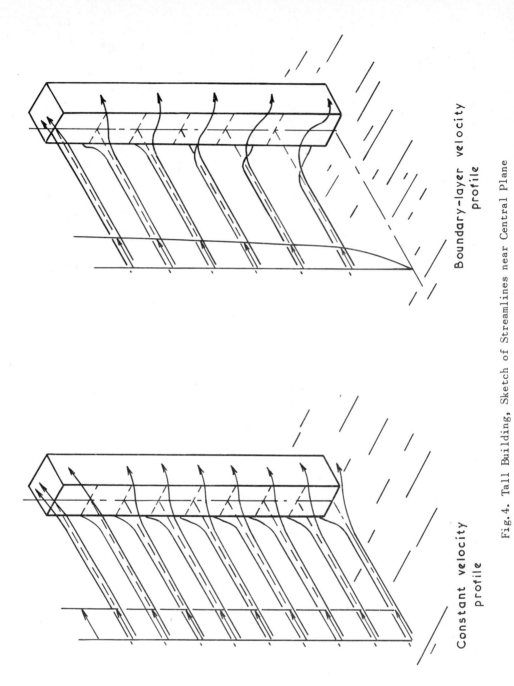

Fig. 4. Tall Building, Sketch of Streamlines near Central Plane

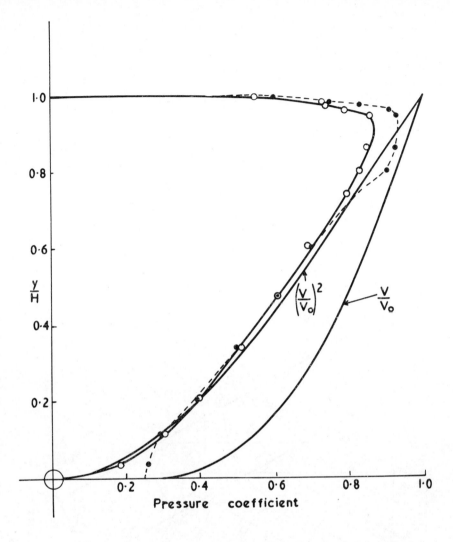

Fig. 5. Tall Building, Pressures on Front Centre Line, $\alpha = 0°$

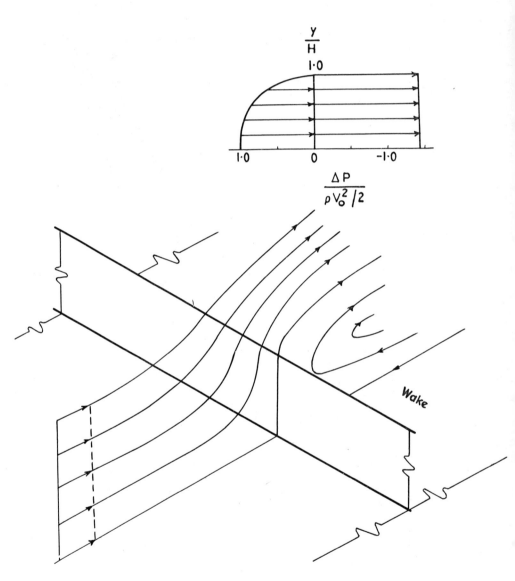

Fig.6. Flow Pattern and Pressure Distribution - Very Long Wall in a Constant Velocity Field

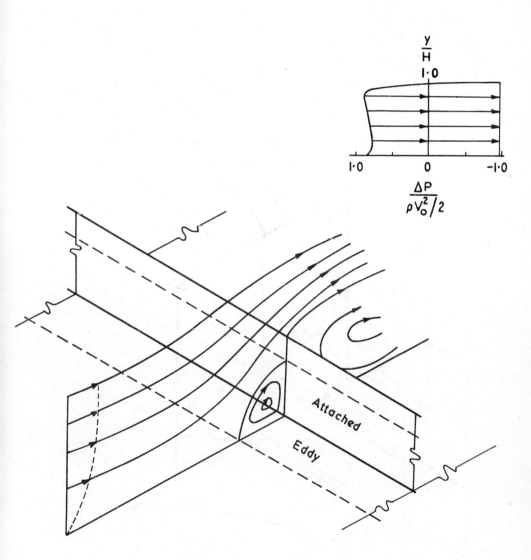

Fig.7. Flow Pattern and Pressure Distribution - Very Long Wall in a Boundary-Layer Velocity Field

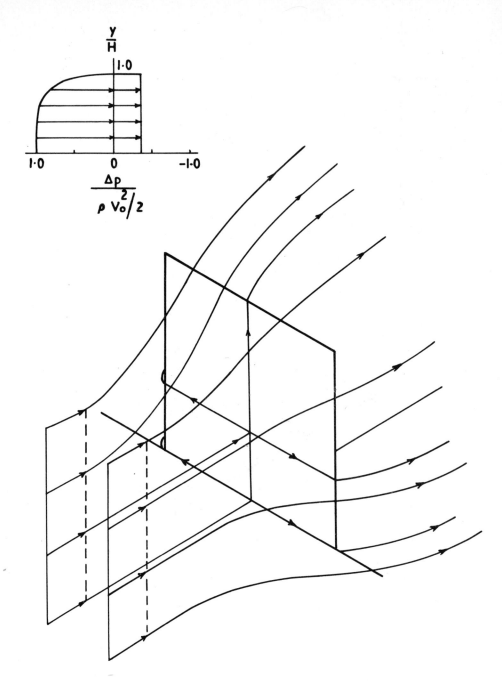

Fig. 8. Flow Pattern and Centre-line Pressure Distribution - Wall of Height: Width = 1:1, in a Constant Velocity Field

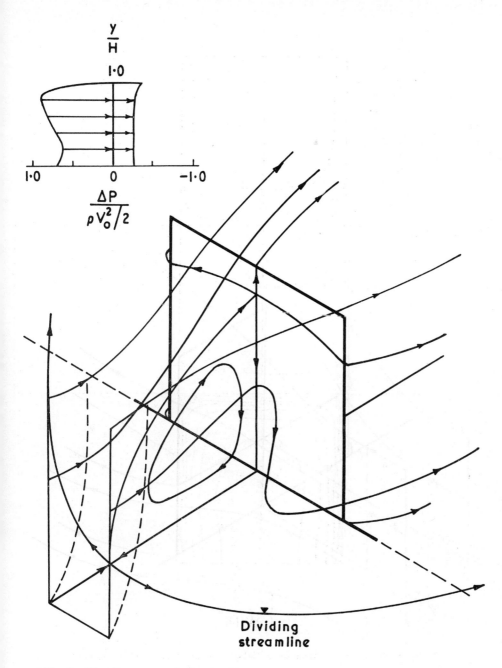

Fig.9. Flow Pattern and Centre Line Pressure Distribution - Wall of Height: Width = 1:1, in a Boundary-Layer Velocity Field

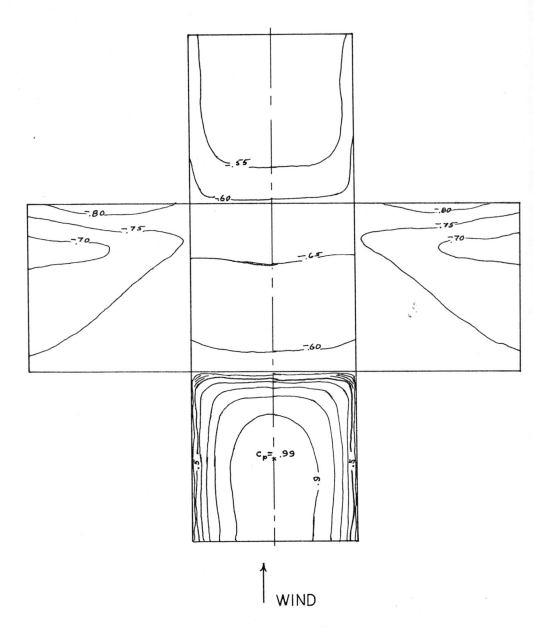

Fig.10. Pressure Distribution on a Cube in a Constant Velocity Field

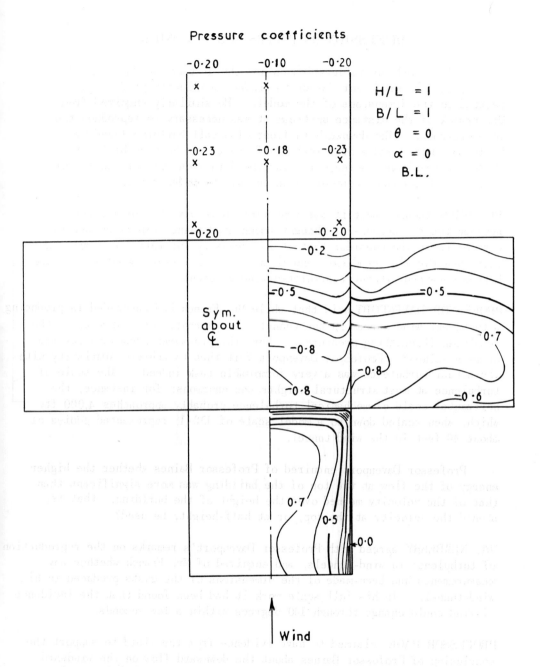

Fig.11. Pressure Distribution on a Cube in a Varied or Boundary-Layer Velocity Field

DISCUSSION ON PAPERS NOS. 15 AND 6

MR. WISE asked for information on how the distance upstream of the screen, used by Professor Baines to produce a velocity profile, was related to the dimensions of the model. He similarly enquired from Mr. Franck of the distance upstream it was necessary to reproduce the local terrain. The downwash in front of a tall building found by Professor Baines had a practical manifestation in Notting Hill Gate where the resultant increase in wind speed in front of a recently built tall block had caused serious inconvenience to pedestrians.

DR. PRIS thought that it was important to resolve the differences between the two schools of thought among wind-tunnel experimentalists; between those who consider that it is necessary for reliable results to reproduce the wind gradient, and those who believe that such results are better obtained without a simulated wind gradient.

PROFESSOR DAVENPORT said that while Mr. Franck had succeeded in producing a turbulent wind stream in the tunnel, he apparently did not measure the resulting fluctuating pressures. Were the pressures presented maximum or mean values? Professor Davenport felt that to achieve similarity with the actual turbulence was a very formidable task indeed. The scale of turbulence at most structural heights was enormous; for instance, the horizontal scale of mechanical turbulence probably approaches 4,000 ft. which, when scaled down to a model scale of 100/1, represented eddies of about 40 feet in the wind tunnel.

Professor Davenport enquired of Professor Baines whether the higher energy of the flow at the top of the building was more significant than that of the velocity meaned over the height of the building; that is, should the velocity at the top, or at half-height, be used?

MR. NEWBERRY agreed with Professor Davenport's remarks on the reproduction of turbulence in wind-tunnels, and enquired of Mr. Franck whether any measurements had been made of the dimensions of the gusts produced in his wind-tunnel. In his full-scale work it had been found that the incidence of wind could change through 140 degrees within a few seconds.

PROFESSOR PAGE claimed to have evidence from the field to support the conclusion of Professor Baines about the downward flow on the windward faces of tall buildings, and he supported this view with illustrations on slides of several incidents which occurred in the Sheffield area during the gales of February, 1962.

MR. FRANCK (in reply) said that the wind tunnel in Copenhagen had an experimental section 7 metres in length and all of this was fitted with the appropriate roughness coating. The boundary layer thickness produced by the tunnel surface should be twice as thick as the height of the model. No measurements of fluctuating pressures due to the turbulence produced in the wind tunnel had yet been attempted.

PROFESSOR BAINES (in reply) said that the distance upstream of the curved screen he used for producing velocity gradient in the wind tunnel was quite short; otherwise the effects of the smoothness of the tunnel floor became evident. In reply to Dr. Pris' question on the necessity for a velocity gradient he said that it depended on the dimensions of the structure under test. For example, the effect of a velocity gradient on a skyscraper was to reduce the load on the roof by as much as a half. On a low building, say four or five storeys, the velocity distribution had no effect on the roof suction. Professor Baines agreed with Mr. Franck that the relevant wind speed was that at the top of the building. The wake pressure is related to the speed at the top of the building. The higher the building the higher the suction not only at the top of the building but also lower down to ground level. On the problem of interference effects on groups of buildings he had found that certain spacings of buildings could produce greater than normal suctions and pressures, but the effects of velocity gradient on these proximity effects was extremely small.

PAPER 5

PREPARATION DES ESSAIS SUR MAQUETTES DE BATIMENTS AU LABORATOIRE AERODYNAMIQUE ET APPLICATIONS A LA VRAIE GRANDEUR

by

DR. RAYMOND PRIS
Ingénieur, E.C.P.

PREPARATION DES ESSAIS SUR MAQUETTES DE BATIMENTS AU LABORATOIRE AERODYNAMIQUE ET APPLICATIONS A LA VRAIE GRANDEUR

by

RAYMOND PRIS

(Paris, France)

AVANT PROPOS

DE nombreux essais ont été éxécutés ces dernières années au laboratoire Eiffel à Paris, en vue de préparer la deuxième edition des Règles Neige-Vent applicable à la Construction en France. Le caractére systématique de ces recherches entreprises en accord avec les membres de la Commission de rédaction présidée par M. Esquillan, a rendu plus apparent les divergences déjà constatées entre les valeurs numériques obtenues sur une même maquette par divers laboratoires pour certains coefficients de pression ou coefficients résultants. Les dispositions adoptées pour le montage des maquettes peuvent, en effet, influer sur les écoulements aérodynamiques.

Ce fait a été déjà signalé par MM. Irminger et Nøkkentved [1], sans que les résultats éveillent l'attention des expérimentateurs. Aussi, une étude spéciale relative à la préparation des essais s'est révélée indispensable dans le but de compléter cette étude. Les divergences entre laboratoires étant supprimées, l'application à la vraie grandeur des résultats obtenus en soufflerie s'effectuera d'une façon plus précise.

La présente communication ne fait donc état que des essais éxécutés dans ce but précis; ceux éxécutés pour la préparation des Règles ont été intégralement publiés dans les 'Annales de l'Institut Technique du Bâtiment des Travaux Publics.'

1. PREPARATION DES ESSAIS SUR MAQUETTE

1 - Essais caractéristiques

1° - Pression du vent sur les bâtiments [1]

La zône la plus sensible à la vitesse et à la direction du courant local est située sur les toitures à deux versants de pente comprise entre 10° et 20°, immédiatement en arrière de la rive amont pour un vent normal à la façade correspondante, *fig.1*. Les pressions relevées en ce point peuvent être prises comme critère pour la comparaison des essais effectués avec des montages différents.

Le montage utilisé par MM. Irminger et Nøkkentved est représenté *fig.2*; les maquettes, dont la hauteur est de 50 mm, sont posées sur la paroi de la soufflerie considérée en tant que sol, à 1,5 m environ de l'entrée du collecteur. Les pressions ont été mesurées au point a, *fig.1*, pour différentes rugosités de cette paroi recouverte de toiles abrasives.

Le coefficient de pression a varié de $c = -1$ pour une faible rugosité à $c = 0,1$ pour une forte rugosité. Ce resultat provient de la variation de la direction locale du courant, *fig.3*, compte tenu de l'inclinaison du versant égale à 20° dans le cas présent. L'ascendance diminue ainsi que la dépression quand la rugosité augmente.

2° - Essais sur modèles et en vraie grandeur dans le vent naturel [2].

Le montage est analogue à celui indiqué ci-dessus; la maquette avait 80 mm de hauteur et la pente des versants était de 6°. Les coefficients de pression c au point a évoluent comme il est indiqué, *fig.4*, en fonction d'un paramètre dont la valeur augmente avec la rugosité de la paroi de la soufflerie.

La dépression est maximum en valeur absolue pour une certaine rugosité; les valeurs extrêmes sont ici, $c = -1,3$ et $c = -0,8$.

3° - Préparation et montage des maquettes en soufflerie [3].

Ces essais ont été éxécutés au laboratoire Eiffel. Contrairement aux précédents, la maquette n'était pas posée sur la paroi mais sur une tôle unie formant sol de 1,40 m de longueur, $fig.5$. Les maquettes étaient constituées par des parallélépipèdes allongés dont la hauteur variait de 30 à 150 mm; la distance X de la face au vent normale au courant au bord avant de la plaque était comprise entre 0 et 700 mm. Le coefficient de trainée ou coefficient résultant a été mesuré en fonction de X; les courbes de la $fig.5$ montrent un évolution importante. Après un maximum obtenu pour une de $X = X_1$, la résistance diminue d'une façon continue.

2 - Interaction couche limite - maquette en soufflerie.
Comparaison couche limite, gradient du vent en vraie grandeur.

1° - Couche-limite sur plaque-sol. Interaction [3].

Les vitesses moyennes ont été relevées dans la couche limite à 80 cm en arrière du bord avant de trois plaques constituées de matières différentes, contreplaqué rugueux, tôle unie, plastique très lisse, $fig.6$. La vitesse de la soufflerie était voisine de 20 m/sec. On voit, par exemple, que la vitesse de 16 m/sec était obtenue à 21 mm au dessus de la plaque en contreplaqué et à 6 mm au dessus des deux autres plaques; à cet écart correspond une différence importante pour la perte d'énergie cinétique. En ce qui regarde la pression, on sait qu'elle n'est pas influencée par le frottement dans la couche limite. Le gradient des vitesses **au dessus de la plaque** $\Delta V/\Delta Z$ est très élevé.

En amont d'une maquette, la vitesse dans cette couche subira une nouvelle réduction correspondant à une augmentation de pression en accord avec l'égalité de Bernouilli; la vitesse sera nulle à proximité de la plaque et les trajectoires seront déviées; il y aura décollement au point A, $fig.7$. Les trajectoires qui, en fluide parfait sans frottement, suivraient la plaque sol, donnant la valeur c = 1 au coefficient de pression au pied de la façade, seront déviées comme il est indiqué $fig.7$ et 8 amenant la formation d'un tourbillon de pied de façade à axe horizontal. Les pressions seront modifiées, principalement sur la façade au vent et sur la toiture. L'étendue et l'intensité de ce tourbillon paraissent hors de proportion avec l'épaisseur de la couche limite dont la vitesse est réduite de 20 m à 14 m sur une épaisseur Z égale a 3 mm $fig.6$, alors que la hauteur de la maquette était de 150 mm et le diamètre du tourbillon environ la moitié, 75 mm.

2° - Vrai grandeur - Gradient du vent.

Les couches atmosphériques s'écoulant près du sol sur des distances évaluées en centaines ou milliers de kilomètres en partant de zones en hautes pressions, ont leurs vitesses diminuées sur une grande hauteur; il en résulte une valeur du gradient des vitesses $\Delta V/\Delta Z$ au sol beaucoup plus faible que celui correspondant à une couche limite. Les courbes inscrites *fig.9* [2] proviennent de mesures effectuées en vent naturel; on a admis V = 20 m/sec pour Z = 100 m. Les gradients au sol et en soufflerie peuvent être comparés *fig.10*.

3° - Maquette et vraie grandeur [3].

Les vitesses du vent en amont d'un bâtiment de 15 m de hauteur relevées sur la courbe 'sol naturel', *fig.9*, sont reportées *fig.11*. Il est évident que les pressions mesurées sur le bâtiment représenté sur cette figure et sur la maquette représentée *fig.7 et 8* diffèrent totalement sous l'influence des gradients au sol. Ce résultat a une grande importance car il n'existe pas d'effet couche limite ni de tourbillon de pied de façade sur un bâtiment en vraie grandeur.

Un montage correct pour les maquettes ne devra donc pas comporter de plaque-sol avant susceptible d'établir au niveau de cette plaque un courant avec gradient des vitesses élevé, analogue à celui existant dans une couche limite.

La répartition des vitesses en hauteur, au dessus de la couche limite, doit faire l'objet d'une étude séparée (p. 7).

3 - Montage avec maquette-image.

1° - Définition de l'élancement.

l_1, l_2 et l_3 étant les désignations des trois côtés d'un bâtiment parallélépipédique, le vent étant normal à la face l_1- l_2 et la longueur l_1 normale au sol, on a:

Élancement = $\dfrac{l_1}{l_2} \gtrless 1$. *fig.12 et 13*

2° - Montage des maquettes, $E > 1$. *fig. 12*

L'écoulement ou sillage en arrière d'un tel volume est co........ le décollement le long de l'arête la plus longue l_1, normale a.......... sait qu'il se forme une série de tourbillons alternés dits de Karman;....

trajectoires traversent le plan de symétrie parallèle au courant, *fig.14, 15 et 16*.

La maquette sera montée en bordure de la plaque-sol laquelle s'étendra latéralement et en aval sur une longueur suffisante pour établir l'intensité tourbillonnaire normale autour de la maquette, *fig.17*. De plus, une maquette image sera prévue afin de rendre l'écoulement général parallèle à la plaque-sol. Celle-ci peut être supprimée sans apporter de changement dans l'écoulement.

3° - Montage des maquettes, $E < 1$. *fig.18*

La nature du sillage est commandée par l'arête l_2 parallèle au sol. Une série de tourbillons alternés ne peut s'établir puisque les trajectoires doivent être parallèles à la plaque sol. Il se forme un tourbillon stationnaire avec courant de retour au contact de la maquette, *fig.18 et 19*. Si la plaque sol est supprimée des tourbillons alternés se formeront correspondant à un élancement $E > 1$.

L'étude d'un bâtiment ayant un élancement $E < 1$ ne peut se faire qu'en présence d'une plaque-sol latérale et arrière.

4 - Montage sans maquette image

1° - Principe du montage

Le montage précédent remplit toutes les conditions pour que les résultats obtenus soient considérés comme des critères en vue de la comparaisons entre laboratoires. Il oblige à prévoir double maquette et à exécuter un montage souvent difficile. On a cherché à supprimer la maquette image en partant des résultats signalés *fig.5* et du tableau joint, pour des bâtiments avec $E < 1$. On observe que pour $X = X_1$, valeur sensiblement égale à la plus petite dimension du maître couple, le coefficient c_t atteint la valeur prise comme critère, avec maquette image et $X = 0$. Ce fait s'explique en remarquant que pour l'écartement X_1 l'effet image n'est plus sensible et que cette longueur est trop faible pour créer une couche limite importante. Un tel montage pourra donc être adopté pour des bâtiments parallélépipédiques et à plus forte raison **pour** des constructions cylindriques à résistance moindre.

2° - Vérification du montage - Mesure des pressions sur toitures à deux versants et en voute, montage standard et montage simplifié.

Il a été signalé p.4, que les toitures étaient particulièrement influencées par les caractéristiques du courant local. Pour cette raison, les coefficients de pression ont été établis sur des maquettes ayant les caractéristiques indiquées *fig.20, 21, 22 et 23*. Les lettres a, b et h remplacent l_3, l_2 et l_1. Les pentes des versants étaient 0°, 10°, 20° et la flèche relative 1/5.

Les valeurs des pressions moyennes sur chaque génératrice parallèle au faitage sont inscrites en montage standard et simplfié, ainsi que les moyennes générales sur chaque versant [3].

On voit que, sauf pour les courbes marquées E et F, lesquelles sont décalées d'un faible pourcentage, il y a accord entre les valeurs ponctuelles et moyennes correspondant aux deux montages.

Il y a lieu de penser que les coefficients sur la face au vent seront légèrement reduits.

Conclusion

Des essais complémentaires seraient à éxécuter sur de nouvelles maquettes de formes différentes. On peut néanmoins assurer que, en adoptant le montage simplifié, les divergences entre laboratoires seront notablement réduites.

2. APPLICATION A LA VRAIE GRANDEUR DES RESULTATS OBTENUS EN LABORATOIRE

1 - Répartition des vitesses au dessus du sol et en soufflerie

1° - Position du problème

Il est reconnu:

> que les vitesses moyennes du vent augmentent avec la hauteur,
>
> que la turbulence du vent augmente à proximité du sol et peut exceptionnellement inverser le sens de l'évolution des vitesses [4].

Il y a lieu de savoir s'il est nécessaire de reproduire en soufflerie une répartition semblable de la vitesse du courant, compte tenu de l'échelle des maquettes, pour que les coefficients de pression soient applicables sans correction à la vraie grandeur, étant entendu que les règles de similitude sont observées.

2° - Effets du gradient du vent sur maquette à grande échelle

(a) Préparation de l'essai [2]

Une maquette de bâtiment de 80 mm de hauteur a été montée sur la paroi de la soufflerie, *fig.2*, et la rugosité de cette paroi a été modifiée de façon à reproduire à l'échelle de la maquette (1/20), le diagramme des vitesses du vent relevé au dessus du sol à un moment donné. La hauteur du batiment était de 1 m 60 et celle de la maquette de 8 cm.

(b) Calcul des coefficients de pression

Il doit être tenu compte des observations suivantes pour le choix de la vitesse de référence utilisée pour le calcul des coefficients:

la vitesse moyenne évolue d'une façon continue avec la hauteur; il y a lieu de déterminer celle à laquelle sera faite le relevé de la vitesse.

les turbulences en vent naturel et en soufflerie sont probablement différentes pour deux niveaux correspondants. Or, la vitesse moyenne établie après examen d'un diagramme en vent turbulent dépend de la nature de l'enregistrement et, en particulier, de la vitesse de déroulement du cylindre. Les diagrammes suivants sont donnés en exemple; ils correspondent à des déroulements de 1/276 mm/sec pour les postes de la Météorologique Nationale, *fig.24*, et 1 mm/sec pour le relevé de la turbulence, *fig.25*. L'étude de ces diagrammes peut amener à des conclusions non identiques.

les moyennes relevées sur les diagrammes doivent être établies en fonction des carrés des vitesses, homogènes à des pressions, et non en fonction des vitesses elles-mêmes.

(c) Résultats

Pour la valeur modérée de l'échelle adoptée et en fonction des vitesses choisies comme références au laboratoire et en vraie grandeur, on obtient[2] sur les façades et les versants des coefficients de pression comparables.

3° - Effets du gradient du vent sur maquettes à échelle réduite

   (a) Modification du diagramme des vitesses
       Apparition d'une couche limite

L'échelle adoptée pour la préparation des maquettes est très souvent égale à 1/100 et même à 1/200 de la vraie grandeur. Une telle réduction des hauteurs appliquée au diagramme des vitesses modifie profondément le gradient ainsi que les échanges d'énergie entre couches d'air voisines. Reprenant les *fig.9 et 10*, et réduisant au 1/100 ou 1/200 les abcisses Z du diagramme 'Sol naturel', on voit que l'on rétablit les gradients correspondant à ceux obtenus en soufflerie pour une couche limite; un tourbillon de pied de façade prendra naissance, *fig.7 et 8*, et les mesures obtenues au laboratoire ne seront plus applicables sans correction à la vraie grandeur.

L'air ne peut plus être considéré comme un fluide sans viscosité pour les valeurs élevée du gradient et le paramètre de Reynolds doit intervenir, c'est à dire égalité du produit (vitesses x hauteurs) en vraie grandeur et sur modèle, tout au moins dans certaines limites de la vitesse. La similitude des essais n'est plus réalisée par simple réduction de la hauteur (Z).

   (b) Suppression de la couche limite

Un montage théoriquement correct serait réalisé par l'interposition de grillages plus ou moins nombreux montés en amont des maquettes, établissant un diagramme des vitesses semblable à celui correspondant à la vraie grandeur, mais évitant la formation d'une couche limite au niveau du sol, *fig.26*. L'écoulement correspond alors à celui d'un fluide parfait; toutefois, la turbulence fine créée par les grillages pourra agir sur les corps à surfaces arrondies, sphères et cylindres.

   (c) Réalisation d'un gradient des vitesses en soufflerie

Il existe de multiples gradients en vent naturel, même en un lieu donné, dépendant de la direction et de la vitesse; le gradient soufflerie ne pourra donc être déterminé qu'après une étude préalable sur le terrain. Actuellement, les renseignements concernant cette question sont très insuffisants et la réalisation en soufflerie d'une répartition déterminée des vitesses compliquera notablement la préparation de l'essai et augmentera la durée de son exécution.

2 - Choix des vitesses de référence en vraie grandeur et au laboratoire - Calcul des pressions en vraie grandeur

1° - Vraie grandeur

Les vitesses sont données par les Règles administratives en vigueur, en fonction de la hauteur et de la position géographique du lieu de la construction. Des données locales peuvent être prises en considération.

2° - Laboratoire

L'un des montages normalisés ou simplifié décrits p. 6 & 7 étant réalisé, les coefficients de **pression** seront établis d'après les pressions mesurées sur maquette, en prenant une vitesse de valeur constante, celle de la soufflerie.

3° - Calcul des pressions

En l'absence de l'effet de couche limite, les vitesses en amont d'une maquette ont une valeur constante (V_s) en soufflerie alors que, en réalité, ces vitesses sont (V_z) variable, *fig. 27*. On admet que $V_z = V_s$ à une certain hauteur (Z_2) supérieure à celle de la maquette. En un point M de la face amont d'une maquette, on mesurerait les pressions p_s et p_z correspondant à ces deux diagrammes, p_z étant la pression réelle en vraie grandeur.

$$(1) \quad p_s = c_s \frac{\rho V_s^2}{2} \qquad (2) \quad p_z = c_z \frac{\rho V_z^2}{2} \quad \text{et} \quad p_s > p_z .$$

Si l'on considérait chaque tranche ΔZ isolément, p_s et p_z seraient alors proportionnels à $(V_s)^2$ et $(V_z)^2$, et $c_s = c_z$. Il en résulte la règle suivante:

Les pressions en vraie grandeur sont obtenues en partant des coefficients donnés par le laboratoire avec V_s constant et en considérant les vitesses réelles V_z variables.

En réalité, il y a interaction des couches fluides, et les pressions ont tendance à s'uniformiser; on peut penser que cette correction est de deuxième ordre. Des essais permettraient de s'en assurer.

3 - <u>Similitude</u>

    1° - Bâtiments limités par des surfaces se coupant suivant des arêtes vives.

Les essais étant éxécutés après montage correct de la maquette, celle-ci doit satisfaire à certaines conditions pour que les coefficients de pression soient identiques au point homologue en vraie grandeur.

L'effet de similitude dépend de la viscosité, générateur d'un sillage en dépression, prenant naissance sur la maquette suivant des lignes de décollement qui s'établissent comme il a été indiqué p.6 & 7. Le décollement a lieu obligatoirement ici le long de la même arête, sur maquette et en vraie grandeur. Il en résulte la similitude des écoulements et des pressions et la constance du coefficient de pression (droites 14 et 15 de la *fig.28*).

    2° - Bâtiments limités par des surfaces courbes [4].

Sur la *fig.28* sont portées les valeurs du coefficient de résistance c_t pour des cylindres présentant différents états de surface: rugosité, nervuration, dont l'importance égale celle du nombre de Reynolds. Alors que pour un cylindre poli, le coefficient est encore évolutif pour le produit vitesse x diamètre égale à 30 m^2/sec, la similitude est réalisée à partir de V x D = 4, pour des cylindres recouverts de toiles abrasives, (courbes 4, 5, 6). Ce fait est du à la disparition des fortes dépressions latérales (*fig.29*, cylindre poli) en présence d'une rugosité ou d'une nervuration (*fig.30*, nervures égales à 1/300 du diamètre). En construction, les surfaces ne sont jamais polies; néanmoins, le produit V x D maquette ne doit pas descendre au dessous d'une certaine valeur, fonction de l'état de surface.

Il est difficile de définir la rugosité d'une construction; elle évolue d'ailleurs avec le temps; en attendant que des études soient entreprises en vraie grandeur, il semble que le classement en forte, moyenne, faible rugosité puisse être adopté.

3. CONCLUSIONS GENERALES

1 - Les conditions de similitude entre maquette et vraie grandeur ne seront réalisées que si la vitesse du courant en soufflerie n'évolue pas trop rapidement en amont des maquettes dans un plan normal au courant sous l'effet de la viscosité. Un gradient analogue à celui d'une couche limite sur plaque plane ne doit pas être formé.

2 - Cette condition amène à prévoir un montage dit normalisé comprenant sol latéral et arrière, avec montage de la maquette et d'une maquette image au bord avant de la plaque-sol.

3 - Un montage simplifié donnant une précision suffisante pour les mesures permet de supprimer la maquette image, à condition de reculer la maquette principale d'une longueur égale, en première approximation, à la plus petite dimension du maître couple.

4 - La transposition sans correction à la vraie grandeur des coefficients de pression obtenus dans les souffleries à vitesse uniforme est justifiée. Le calcul de la pression sera effectué avec une bonne approximation en tenant compte de la valeur locale du vent correspondant à la hauteur au dessus du sol de l'élément considéré.

5 - La similitude des maquettes est toujours réalisée au cours des essais pour les bâtiments comportant des facades se coupant suivant des arêtes vives.

Pour des constructions comportant des surfaces arrondies, le nombre de Reynolds intervient sous la forme du produit V x D out V x L, lequel ne doit pas descendre au dessous d'une certaine valeur fonction de l'état de surface. Celui-ci importe autant que le nombre de Reynolds pour l'établissement de la similitude.

BIBLIOGRAPHIE

1. IRMINGER et NØKKENTVED - Pression du vent sur les constructions - Copenhague, 1930-1936.

2. JENSEN, M. - The Model Law for phenomena in natural wind - Copenhague, Ingenioren, inter. éd., 1958, 2.

3. PRIS, R. - Action du vent sur les bâtiments et constructions. Préparation et montage des maquettes en soufflerie, dans Annales de l'Institut technique du Bâtiment, février 1962.

4. Centre expérimental de Recherches et d'études du bâtiment et des Travaux publics. Mesure de l'effet du vent sur les bâtiments de grande envergure. - Paris, 1962.

5. PRIS, R. - Influence de la nervuration et de la rugosité sur la résistance aérodynamique des cylindres, dans Annales de l'Institut technique du Bâtiment, novembre 1960.

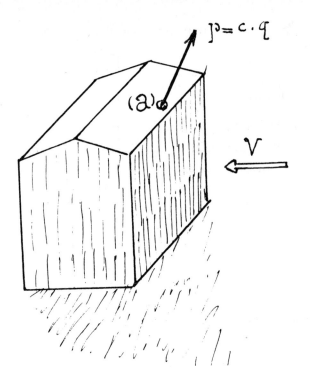

Fig.1

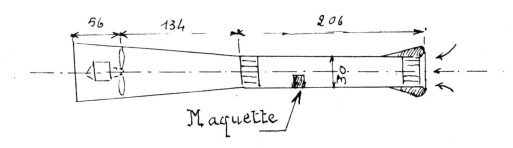

Fig.2

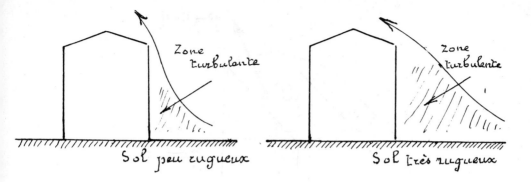

Fig. 3

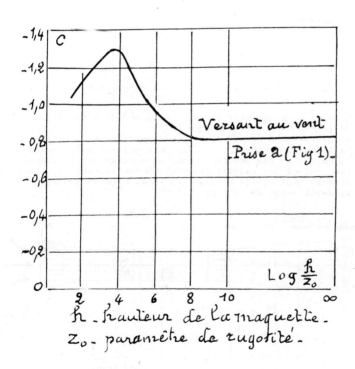

h - hauteur de la maquette.
Z_0 - paramètre de rugosité.

Fig. 4

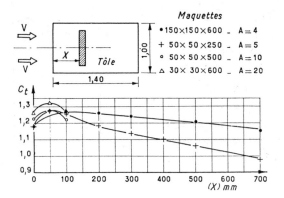

Valeurs comparées de c_t

		150×150 ×600	50×50 ×250	50×50 ×500	30×30 ×600
	MAQUETTES				
Sans sol avant avec image $X = 0$	A	4	5	10	20
	$C_t =$	1,26	1,27	1,20	1,30
Sans image	$X = X_1$	125	50	50	40
	$C_t =$	1,27	1,276	1,27	1,32

Fig.5. Montage sans maquette image. Influence de l'emplacement de la maquette sur la plaque-sol.
X croissant à partir de zéro, le coefficient résultant passe par un maximum; il diminue ensuite régulièrement. La valeur du maximum correspond au montage dit 'standard' pour lequel les valeurs mesurées du coefficient c_t sont sensiblement égales à celles obtenues avec le montage maquette image et X = 0.

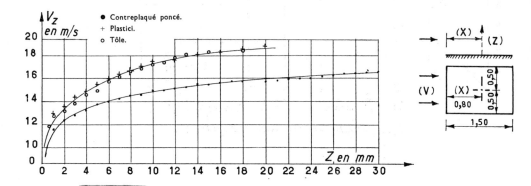

Fig. 6. Relevé des vitesses à l'intérieur de la couche limite formée au contact de différentes surfaces. Son épaisseur caractérise la rugosité de la surface. Après un parcours de 80 cm, son épaisseur minimale est de 20 mm pour une surface lisse (tôle, plastici).

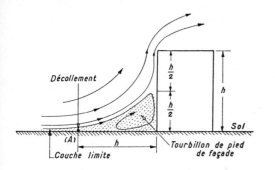
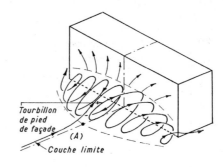

Fig.7   Fig.8

Interaction couche limite-maquette. La surpression existant en amont d'une façade ajoute son effet à celui de la couche-limite pour réduire les vitesses d'écoulement à la surface du sol. Les vitesses peuvent s'annuler à l'intérieur d'une lame d'épaisseur assez importante pour amener un décollement (point A). Il apparait un tourbillon d'axe horizontal en dessous de l'écoulement normal.

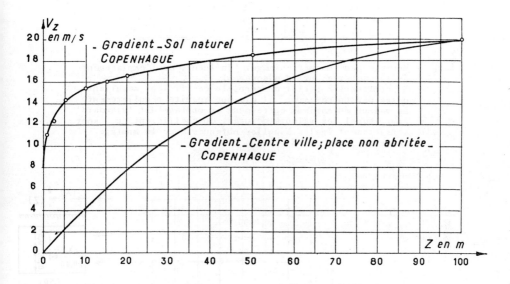

Fig.9. Gradient des vitesses du vent relevé au-dessus d'un sol cultivé, sans dénivellations importantes. On a admis une vitesse égale a 20 m/s à la hauteur $z = 100$m.

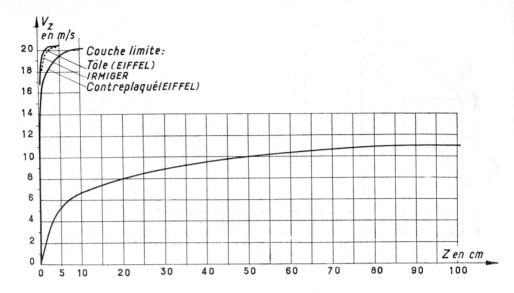

Fig.10. Comparaison des gradients, vent naturel et courant de couche limite sur plaque-sol. Les évolutions sont totalement différentes. Le gradient naturel est beaucoup moins accentué que celui obtenu sur plaque-sol (hauteurs z en centimètres).

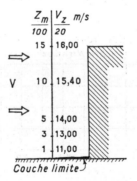

Fig.11. Gradient des vitesses en vent naturel, fig.9, devant la façade d'un bâtiment de 15 m de hauteur. L'écart des vitesses au sol et à la rive de la toiture est trop faible pour amener la formation d'un tourbillon de pied de façade, fig.7-8.

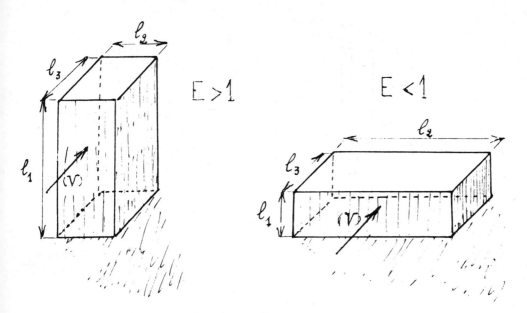

Fig.12 Fig.13

Fig.14 - $wb/v = 250$
Flow round sharp plate of width b

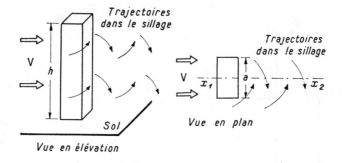

Fig. 15   Fig. 16

Bâtiment élancé. Les trajectoires, à l'intérieur du sillage, correspondent à l'écoulement tourbillonnaire visible sur la *fig.14*.

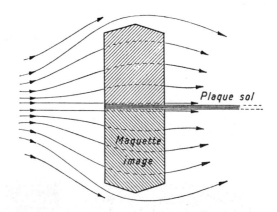

Fig.17. Maquette montée en bordure de plaque-sol, avec maquette image. La symétrie du courant est rétablie. La vitesse de référence pour le vent est celle de la soufflerie. Les mesures seront correctes.

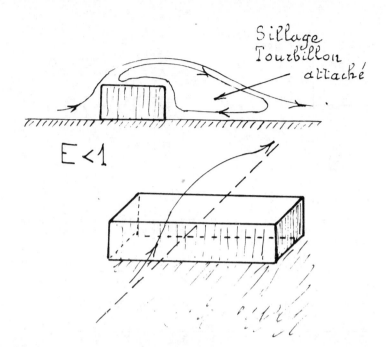

Fig.18

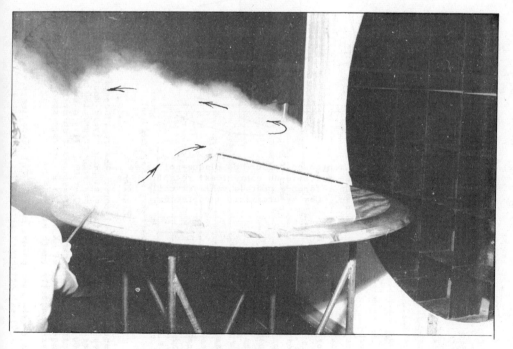

Fig.19

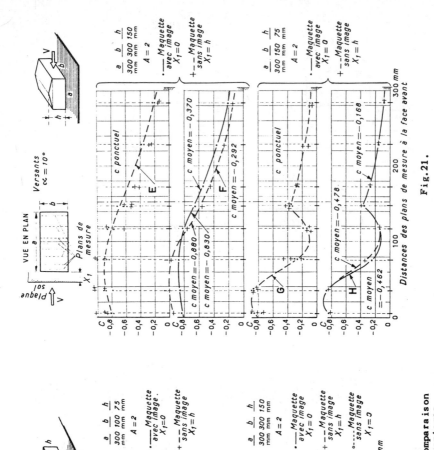

Fig. 20.

Maquette de bâtiment allongé. Mesure et comparaison des pressions sur terrasse et pour divers montages. Les dimensions des terrasses sont en mm: 150 x 300 et 300 x 300 (Courbes A, B, C et D). Il y a bon accord entre les coefficients correspondant au critère (avec maquette image et $X = 0$) et au montage standard (sans maquette image et $X_1 = h$).

La courbe avec points marqués (o) (sans image et $X = 0$) est nettement séparée des précédentes.

Fig. 21.

Maquettes avec toiture à deux versants. Pente $\alpha = 10°$ (Courbes E à H).

Mêmes remarques que pour la maquette avec terrasse. Toutefois, les courbes correspondant aux dimensions $a = b = 300$ mm ne coïncident pas totalement (courbes F). Les causes de cet écart n'ont pas été recherchées.

248

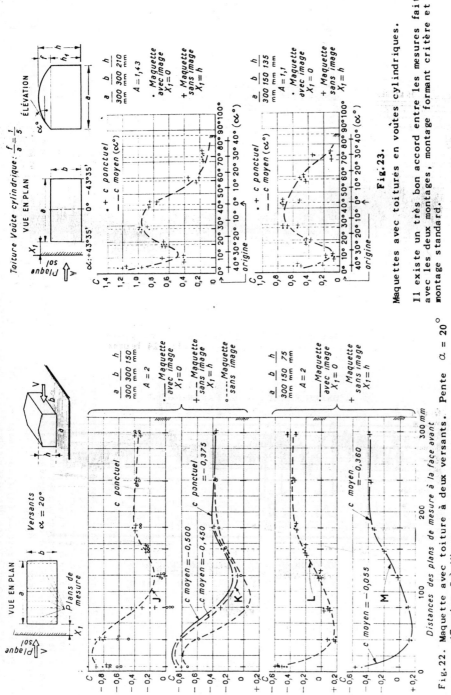

Fig. 22. Maquette avec toiture à deux versants. Pente $\alpha = 20°$ (Courbes J à M).

Mêmes remarques que précédemment (fig. 20). La courbe : maquette sans image ($a = b = 300$ mm) et $X_1 = 0$ est nettement séparée des courbes K.

Fig. 23. Maquettes avec toitures en voûtes cylindriques.

Il existe un très bon accord entre les mesures faites avec les deux montages, montage formant critère et montage standard.

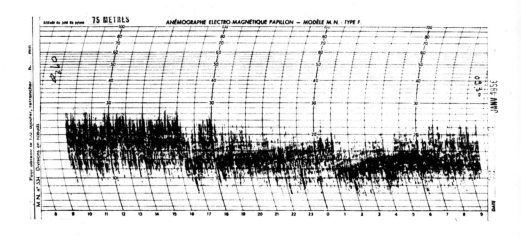

Fig.24. Hauteur: 39 m - Déroulement: 24 heures

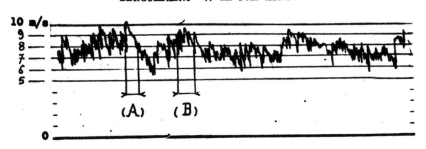

VENT NORD. PLAINE DE LA CRAU

Fig.25

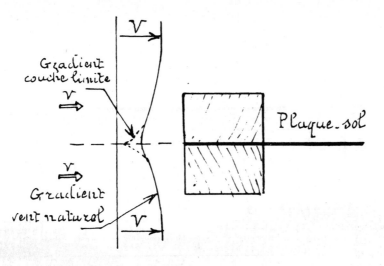

Fig.26

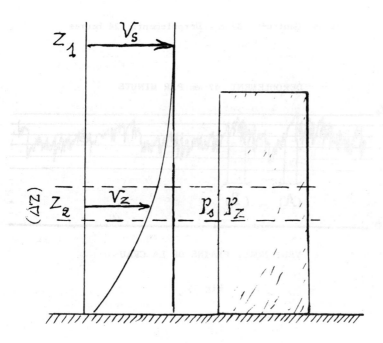

Fig.27

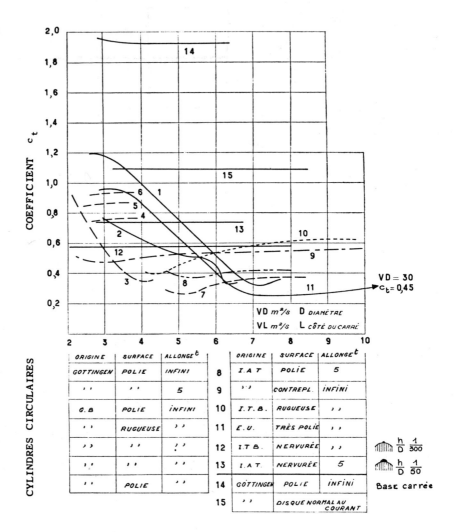

Fig.28. Étude de la similitude. Evolution du coefficient de résistance c_t. Paramètres influant sur la similitude: produit VD m^2/s, turbulence du courant.

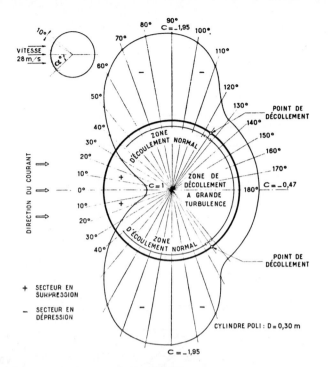

Fig.29

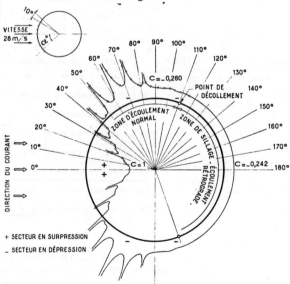

Fig.30

PAPER 22

EXECUTION EN TUNNEL AERODYNAMIQUE D'ESSAIS SUR MAQUETTES DE BATIMENTS EN RAPPORT AVEC LES MESURES FAITES SUR CONSTRUCTION REELLE

by

P. E. COLIN: Training Center for Experimental Aerodynamics

and

R. D'HAVE :   Bureau de Contrôle pour la Sécurité de la Construction en Belgique.

EXECUTION EN TUNNEL AERODYNAMIQUE D'ESSAIS SUR MAQUETTES DE BATIMENTS EN RAPPORT AVEC LES MESURES FAITES SUR CONSTRUCTION REELLE

par

P. E. COLIN and R. D'HAVE

1. INTRODUCTION

LORSQU'IL s'agit d'étudier l'action du vent sur des maquettes de bâtiments, on peut estimer que l'échelle des essais ou le nombre de Reynolds qui y correspond n'auront pas d'influence sur les mesures étant donné qu'il s'agit de constructions à angles vifs pour lesquelles le décollement de la veine d'air se situe aux arêtes de la surface au vent. On serait dès lors tenté de conclure que l'étude sur maquettes ne présente pas de difficultés et que les valeurs des coefficients obtenus peuvent être transposées directement à la vraie grandeur.

Or, si l'on examine les prescriptions qui existent dans différents pays quant aux valeurs à adopter pour les coefficients de pression aérodynamique locale sur les parois de bâtiments, on constate, dans des cas identiques, des divergences importantes. Dans bien des cas, le vent constitue une des sollicitations principales agissant sur la construction et, en vue de répondre au souci toujours plus grand d'économie, il nous est apparu indispensable de mettre au point une méthode d'essai susceptible de fournir des données valables pour les bâtiments réels.

Certains travaux ont déjà été consacrés à ce problème dans plusieurs laboratoires, et il semble qu'il y ait des divergences en ce qui concerne précisément les conditions d'essais à réaliser en soufflerie. Ces conditions sont étroitement liées à l'écoulement qu'il convient de reproduire autour de la maquette en présence d'un plancher représentant le sol et c'est à ce point particulier que l'étude décrite dans ce rapport est consacrée.

Une méthode d'essai a été récomment proposée par l'Institut Technique du Bâtiment et des Travaux Publics, à Paris, dans le cadre de normes à

établir pour la construction. Elle se réfère, comme critère, à la
configuration représentée par la maquette placée avec image au bord
d'attaque de la plaque-sol. Ce cas est inclus dans l'étude que nous avons
effectuée et qui comporte complémentairement aux essais en soufflerie des
mesures sur un bâtiment existant.

Les essais sur maquettes ont été réalisés à la soufflerie à faible
vitesse, à veine libre de 3 m. de diamètre, du Centre de Formation en
Aérodynamique Expérimentale, à Rhode St. Genèse.

2. NOTATIONS

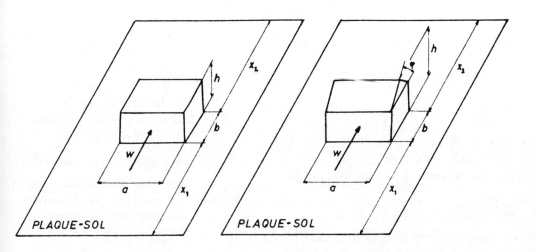

a: dimensions horizontale de la maquette perpendiculairement au vent

b: dimension horizontale de la maquette parallèlement au vent

h: hauteur de la maquette

ϕ: angle d'inclinaison du versant de toiture sur l'horizontale

$\lambda = \frac{h}{a}$: élancement

$\mu = \frac{b}{a}$: profondeur

X_1: distance de la maquette au bord avant de la plaque-sol

X_2: distance de la maquette au bord arrière de la plaque-sol

q: pression dynamique

γ: coefficient de pression locale

C_h: coefficient de trainée

3. ECOULEMENT SUR UNE PLAQUE PLANE DESTINEE A REPRESENTER LE SOL

LE sol est en général représenté en soufflerie par un plancher fixe sur lequel est placé la maquette du bâtiment à étudier. La distance de cette maquette par rapport au bord d'attaque de la plaque-sol peut avoir une influence sensible sur les mesures. C'est cette influence qui a été déterminée sur quelques modèles dans le cadre de la présente recherche. Il était par conséquent nécessaire, avant de procéder aux essais proprement dits de vérifier l'écoulement sur plusieurs types de plaques-sol. Les différents planchers qui ont servi à cette étude préliminaire sont indiqués par les numéros 1 à 3 au tableau A.

Tableau A.

Plaque-sol N°	longueur m	largeur m	épaisseur mm	bord d'attaque
1	1,4	1,4	2	2 mm (cadre support important)
2	1,4	1,4	10	semi-elliptique (longueur 20 mm)
3	1,5	1	12	2 mm sur une longueur de 120 mm
4	3	3	45	semi-circulaire
5	circulaire	Ø 1,2m	18	arrondi vers l'intrados

La pression dynamique a été mesurée dans l'axe longitudinal des plaques à 125 mm au-dessus de celles-ci.

La figure 1 donne les résultats pour la plaque n° 1 munie ou non d'un volet de bord de fuite d'inclinaison variable. Les valeurs portées en ordonnée représentent le rapport de la pression dynamique mesurée au tube de pitot à la pression de référence de la soufflerie. La courbe I fait apparaître des variations assez importantes de la pression dynamique. Ces variations doivent être attribuées à la circulation qui existe autour de la plaque du fait de la sous-structure importante qui la supporte. Les courbes II à V montrent l'influence d'un volet de bord de fuite en fonction de sa dimension et de son inclinaison. Un tel volet constitue un moyen efficace pour améliorer l'écoulement sur une plaque.

Pour éviter l'inconvénient d'une sous-structure et réaliser un plancher suffisamment rigide, la plaque n°2 a été réalisée avec bord d'attaque elliptique.

Les courbes de la figure 2 montrent les répartitions longitudinales de la pression dynamique pour trois incidences de la plaque. On constate que l'augmentation d'épaisseur du bord d'attaque amplifie la zone d'influence de celui-ci, mais la variation maximum de la pression dynamique est moins importante à l'incidence nulle.

Un faible écart d'incidence modifie fortement la répartition de la pression dynamique au-dessus de celle-ci.

La figure 3 montre que sur la plaque n°3 on a obtenue un écoulement satisfaisant grâce à la faible épaisseur du bord d'attaque.

Les plaques 4 et 5 ont été utilisées pour certains essais seulement. La plaque n°4 de grande dimension permet de réaliser des couches limites relativement importantes. La plaque n°5 est habituellement utilisée pour les essais sur bâtiment et a l'avantage de faciliter la mesure des coefficients aérodynamiques pour toutes les directions du vent.

4. MESURE DU COEFFICIENT DE TRAINEE DE MAQUETTES PARALLELEPIPEDIQUES EN PRESENCE D'UNE PLAQUE DE SOL

CINQ maquettes ont été utilisées pour déterminer l'influence de la plaque-sol sur la mesure du coefficient de traînée. Les caractéristiques des maquettes sont données au tableau B.

Tableau B

N° de la maquette	dimension en mm			N° de la plaque-sol
	h	a	b	
1	100	50	25	3
2	100	50	50	3
3	100	50	200	3
4	250	50	25	2
5	300	150	50	3

Les figures 4 à 8 montrent la variation du coefficient de traînée c_h en fonction de la distance de la maquette au bord d'attaque de la plaque-sol, les mesures étant faites avec et sans maquette image.

Sur chacune des figures est également indiquée la valeur de c_h mesurée sur la plaque circulaire n°5. On voit d'après les figures que le c_h des maquettes avec images ne varie que très peu en fonction de la position par rapport au bord d'attaque. Par contre, dans les essais sans maquette image, on constate près du bord d'attaque une variation plus ou moins sensible du coefficient de traînée. Toutefois à une certaine distance les valeurs coincident avec celles des cas correspondants avec image.

5. MESURE DE LA REPARTITION DES PRESSIONS SUR UNE MAQUETTE EN PRESENCE D'UNE PLAQUE-SOL

LA maquette utilisée a les dimensions suivantes:

a = 300 mm., b = 300 mm., h = 150 mm., ϕ = 20°, λ = ½, μ = 1.

Les mesures de pression locale ont été effectuées sur la maquette avec et sans image pour trois distances du bord d'attaque de la plaque-sol: x_1 = 0, h (150 mm) et 450 mm, les prises de pression étant situées sur les faces verticales au vent et sous le vent et sur les deux versants de la toiture, dans trois plans I, II et III (voir schéma au bas de la fig.9). Les résultats sont présentés sous forme de coefficients de pression aux fig. 9 à 12. On constate une concordance relativement bonne pour la maquette avec image en x_1 = 0 et la maquette seule en x_1 = 150 mm. Les courbes sont en général nettement différentes pour x_1 = 450 mm où l'image ne joue pratiquement plus de rôle.

Afin de mieux se rendre compte de la nature de l'écoulement dans le cas de la maquette avec image en $x_1 = 0$ et dans celui de la maquette en $x_1 = 450$ mm, on a eu recours à la visualisation par fumée. Les deux photos de la figure 13 montrent effectivement des différences importantes pour ces deux cas. Pour la maquette à 450 mm, on distingue un tourbillon au pied de la maquette ce qui n'est pas le cas lorsque celle-ci se trouve au bord d'attaque. La visualisation sur la plaque-sol au moyen d'une couche d'huile, contenant de l'oxyde de titane en suspension, montre sur les photos de la figure 14 la structure détaillée de l'ecoulement pariétal autour de la maquette. Le tourbillon est nettement visible dans le cas $x_1 = 450$ mm et est précédé d'une large zone décollée. Les différences remarquées proviennent essentiellement de l'existence sur la plaque-sol d'une couche limite qui décolle devant la maquette en raison du gradient adverse de pression rencontré.

Pour pouvoir porter un jugement sur les conditions d'essai qu'il convient d'adopter, il faut évidemment pouvoir se raccrocher à des résultats obtenus sur bâtiment réel. C'est ce qui nous a amenés à choisir un bâtiment simple, isolé, situé en un endroit bien dégagé sur un sol horizontal et de mettre en parallèle les mesures faites sur ce bâtiment et celles effectuées en soufflerie sur une maquette qui le représente.

6. COMPARAISON DES MESURES DE PRESSIONS LOCALES SUR BATIMENT REEL ET SUR MAQUETTE

LE bâtiment qui a été équipé de prises de pression sur une façade et la toiture est situé à l'aéroport national de Bruxelles. La figure 15 en donne le schéma ainsi que la position des prises de pression numérotées de 1 à 15. Les mesures ont été effectuées par vent du S-O c'est-à-dire pour la direction 0° de la figure 15 avec une tolérance de ± 15° de part et d'autre. En soufflerie, les mesures sur une maquette à l'échelle 1:100 ont été faites dans les mêmes conditions, c'est-à-dire en prenant pour le calcul des coefficients de pressions la moyenne des valeurs obtenues pour des directions du vent de 0°, + 15° et -15°, les différences n'étant d'ailleurs pas grandes.

En ce qui concerne les essais sur bâtiment réel, un relevé de profils de vitesses du vent à une centaine de mètre en amont du bâtiment a également été effectué. Les résultats sont portés à la figure 16 qui donne également le profil de couche limite turbulente selon la loi 1/7 pour deux valeurs de la vitesse à l'extérieur de la couche-limite.

En soufflerie, les mesures ont été faites sur la maquette dans diverses conditions:

- sur la plaque-sol n° 3 avec $x_1 = 0$ et maquette-image
  avec $x_1 = 450$ mm, sans image

- sur la plaque-sol n° 4 - avec reproduction en soufflerie d'un profil de vitesse semblable à celui relevé en place

- sur la plaque-sol n° 5, la maquette étant placée au centre de la plaque.

La figure 17 donne les résultats des mesures pour tous les essais, présentés sous forme de coefficients de pression avec comme référence la pression locale au 2/3 de la hauteur de la façade au vent du bâtiment. L'examen des courbes montre que pour tous les essais sauf pour le cas de la maquette placée avec image au bord d'attaque de la plaque-sol, il y a dans l'ensemble une concordance satisfaisante.

CONCLUSIONS

LES essais qui précèdent se rapportent à des bâtiments de faible élancement et dans ce cas on peut conclure qu'en vue de l'exécution des essais en soufflerie, la meilleure méthode consiste évidemment à reproduire à l'échelle le profil des vitesses relevé à l'endroit où la construction sera édifiée.

Cette condition reste toutefois impossible à satisfaire lorsqu'il s'agit d'établir à priori les normes à utiliser dans le calcul des constructions. Des résultats suffisamment approchés et qui se situent d'ailleurs du côté de la sécurité peuvent être obtenus en plaçant la maquette à une distance suffisante du bord d'attaque de la plaque représentant le sol.

Il semble dans ce cas que l'écoulement en soufflerie résultant du décollement de la couche limite devant l'obstacle que constitue la maquette correspond bien aux conditions réelles.

Dans le cas de constructions élancées, une étude actuellement en cours semble montrer qu'en soufflerie, l'écoulement est fortement influencé par la distance de la maquette au bord de fuite de la plaque-sol. La figure 18 donne la variation du coéfficient de pression au milieu du toit plat de deux maquettes à section carrée, géométriquement semblables, en fonction de la distance de ces maquettes au bord de fuite de la plaque-sol.

Cette distance influence considérablement le sillage. Entre les deux valeurs extrêmes du coéfficient de pression, l'écoulement est très instable ce qui semble montrer l'existence possible de deux régimes d'écoulement. La valeur maximum de la dépression est atteinte et conservée pour des distances au bord de fuite égales ou supérieure à la hauteur de la maquette.

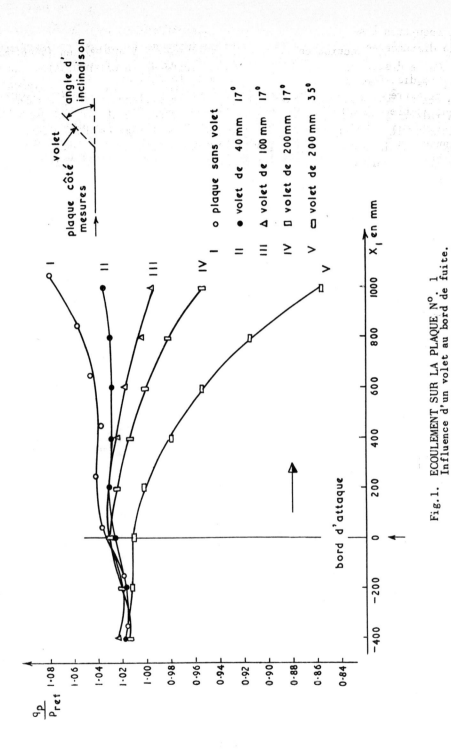

Fig.1. ECOULEMENT SUR LA PLAQUE N°. 1
Influence d'un volet au bord de fuite.

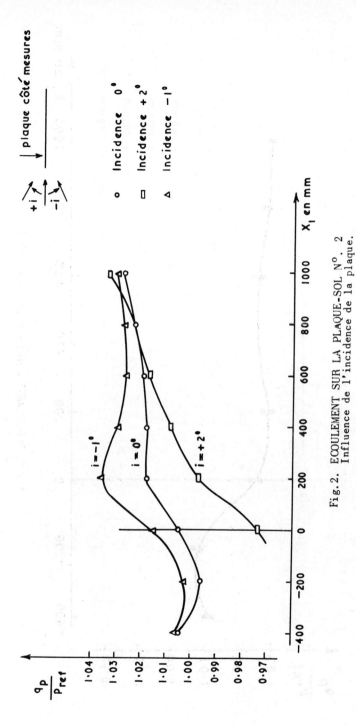

Fig. 2. ECOULEMENT SUR LA PLAQUE-SOL N°. 2
Influence de l'incidence de la plaque.

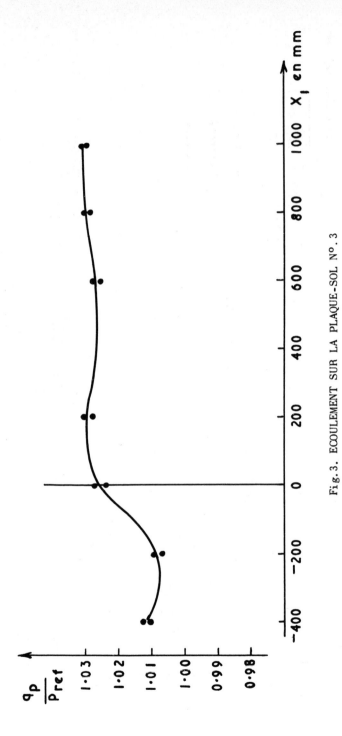

Fig.3. ECOULEMENT SUR LA PLAQUE-SOL N°.3

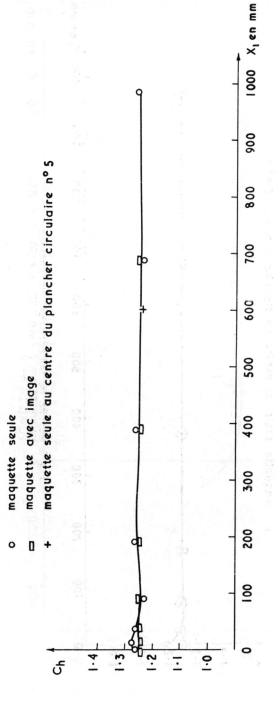

Fig. 4. MESURE DU COEFFICIENT DE TRAINEE
MAQUETTE N°.1 (h = 100, a = 50, b = 25)
PLANCHER N°.3

o maquette seule
□ maquette avec image
+ maquette seule au centre du plancher circulaire n° 5

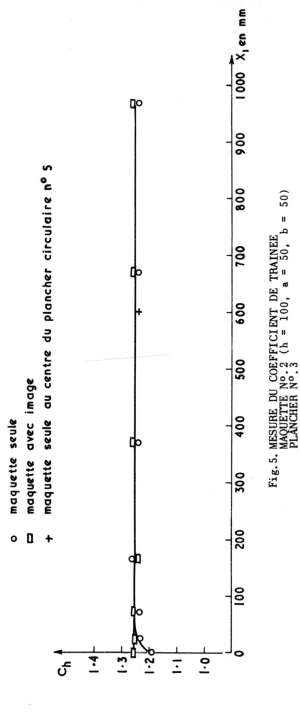

Fig.5. MESURE DU COEFFICIENT DE TRAINEE
MAQUETTE N°.2 (h = 100, a = 50, b = 50)
PLANCHER N°.3

o   maquette seule
□   maquette avec image
+   maquette seule au centre du plancher circulaire n° 5

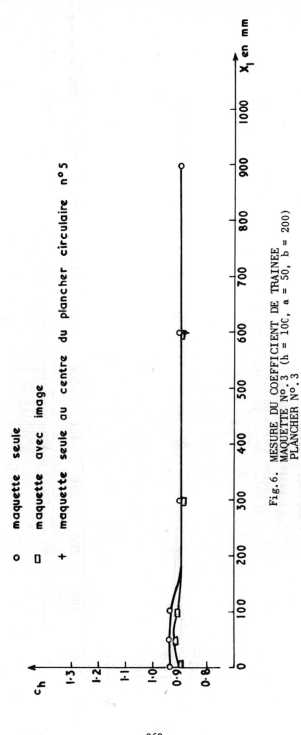

Fig.6. MESURE DU COEFFICIENT DE TRAINEE
MAQUETTE N°.3 (h = 100, a = 50, b = 200)
PLANCHER N°.3

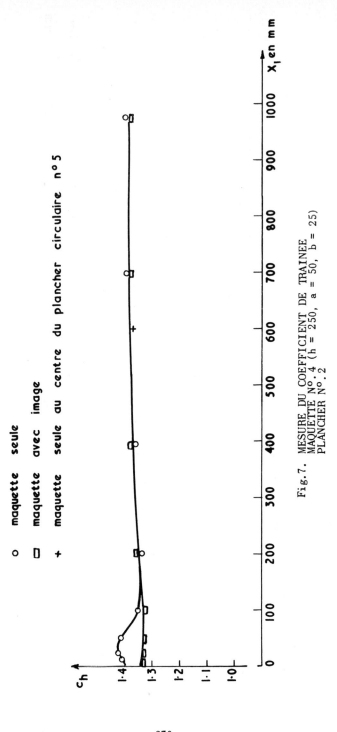

Fig.7. MESURE DU COEFFICIENT DE TRAINEE
MAQUETTE N°.4 (h = 250, a = 50, b = 25)
PLANCHER N°.2

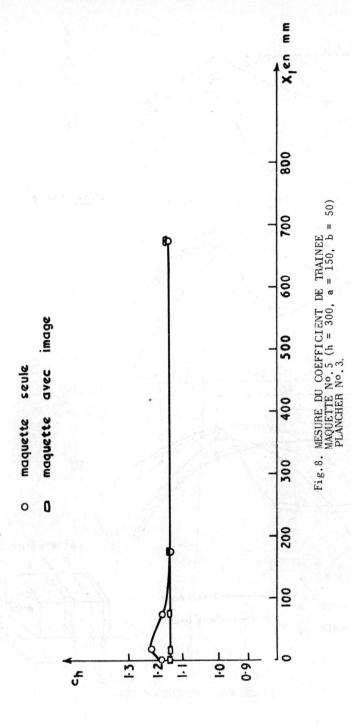

Fig.8. MESURE DU COEFFICIENT DE TRAINEE
MAQUETTE N°. 5 (h = 300, a = 150, b = 50)
PLANCHER N°. 3.

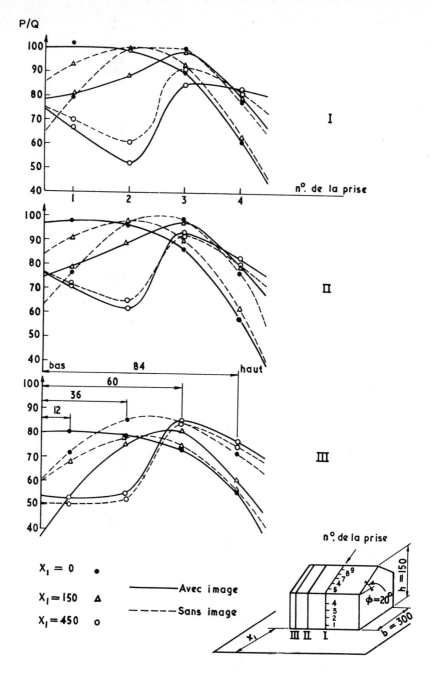

Fig.9. FACE VERTICALE AVANT

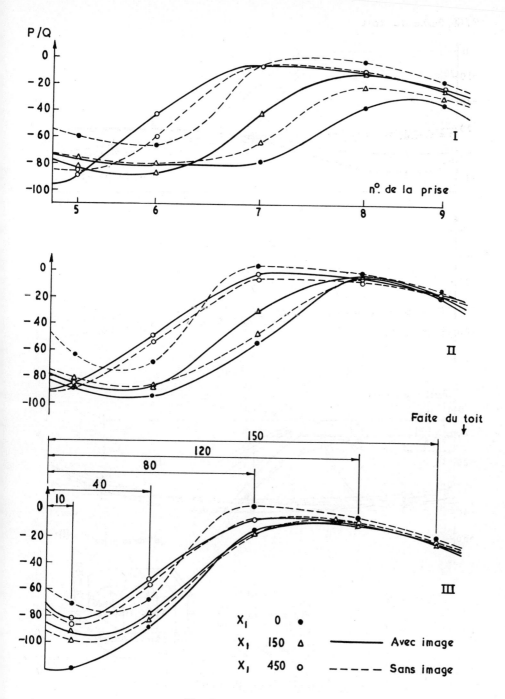

Fig.10. TOITURE AVANT

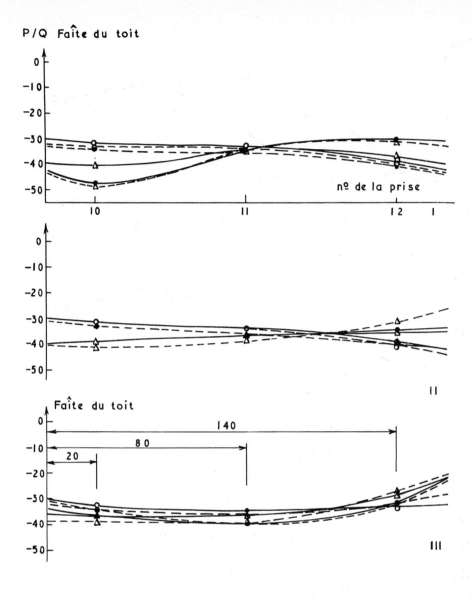

Fig.11. TOITURE ARRIERE

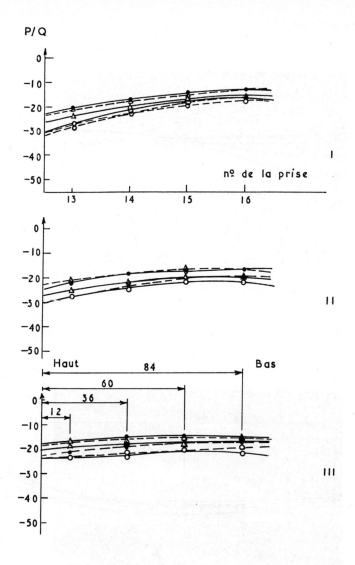

Fig.12. FACE VERTICALE ARRIERE

Fig.13. Maquette avec image en $X_1 = 0$

Fig.13. Maquette sans image en $X_1 = 450$ mm

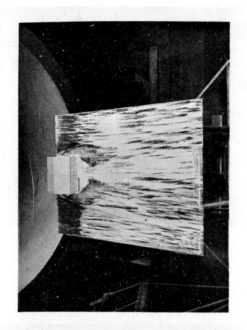

Fig.14. Maquette avec image en $X_1 = 0$

Fig.14. Maquette sans image en $X_1 = 450$ mm

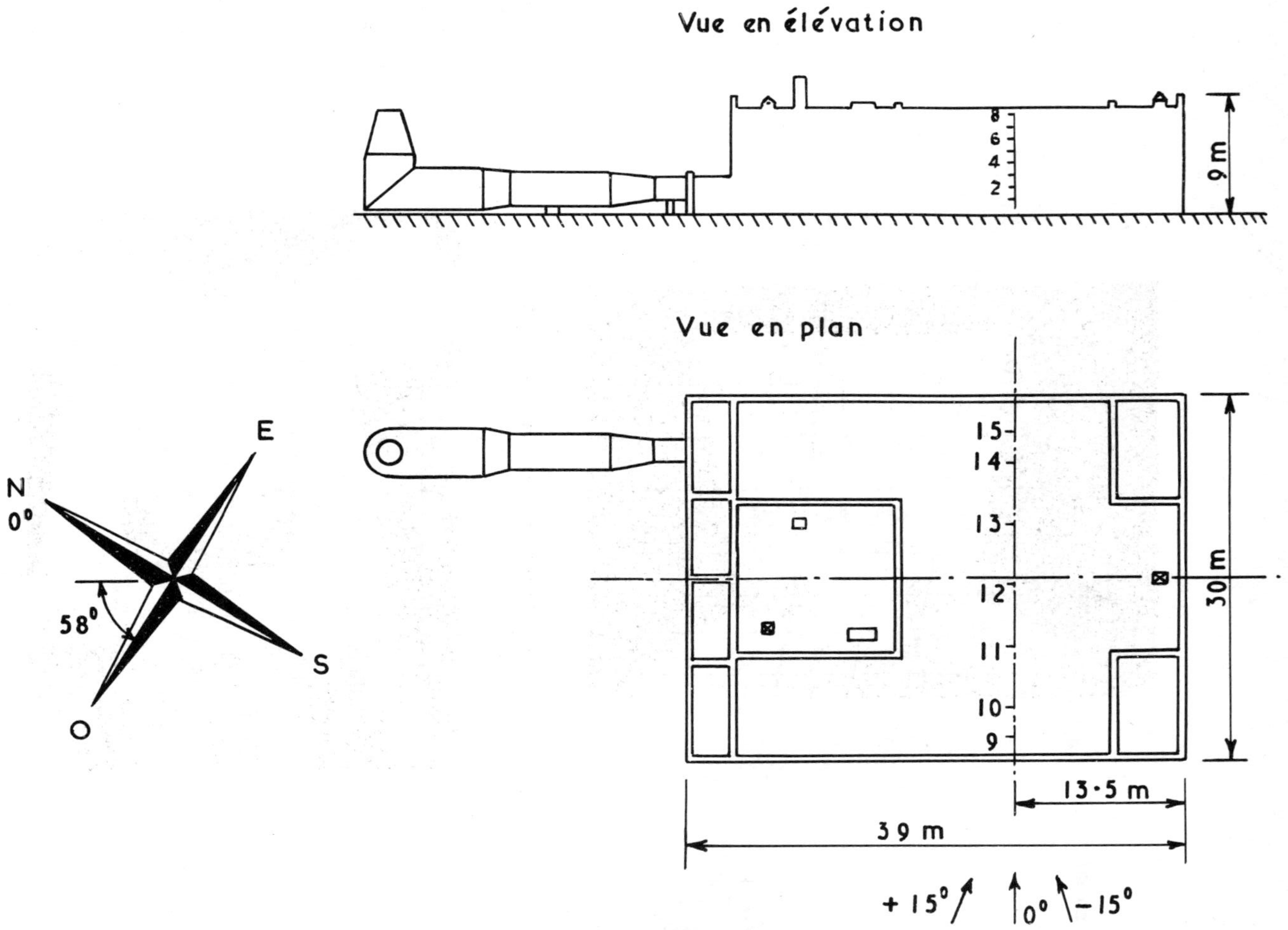

Vue en élévation

Vue en plan

Fig.15. SCHÉMA DU BATIMENT UTILISÉ POUR LES ESSAIS

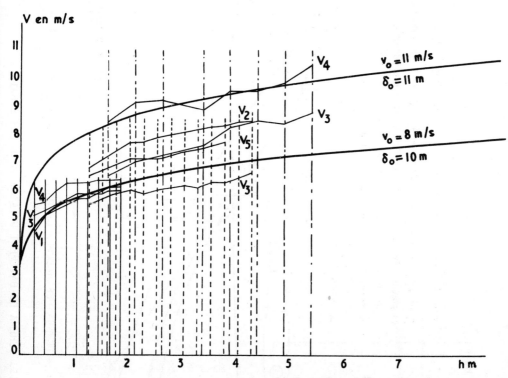

v_o : vitesse à la limite extérieure de la couche limite

δ_o : épaisseur de la couche limite

Fig.16. RELEVE DES VITESSES à 100 M. EN AVANT DU BATIMENT REEL

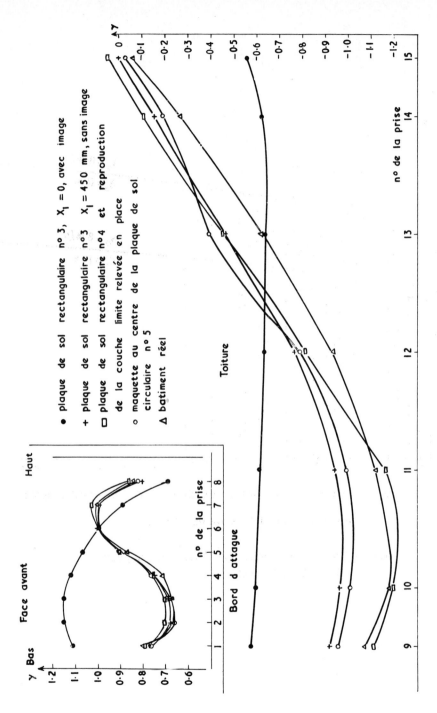

Fig. 17. MOYENNES DES MESURES DE PRESSIONS LOCALES SUR BÂTIMENT RÉEL ET SUR MAQUETTE.

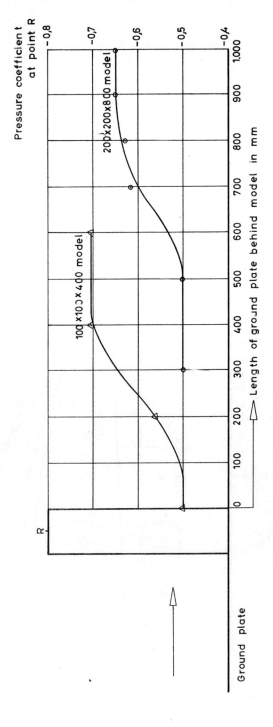

Fig.18. Effect of ground plate length behind model on pressure distribution

PAPER 21

MODEL SIMULATION OF WIND EFFECTS ON STRUCTURES

by

R. E. WHITBREAD
(National Physical Laboratory)

MODEL SIMULATION OF WIND EFFECTS ON STRUCTURES

by

R. E. WHITBREAD
(National Physical Laboratory)

1. INTRODUCTION

THIS paper attempts to set out briefly the similarity requirements which need to be observed in order that the results of tests carried out in a wind tunnel may be used to predict the behaviour of the full-scale prototype in the natural wind. These requirements will vary according to the problem under investigation. To achieve similarity it is necessary to represent in the wind tunnel the relevant physical properties of both the structure and of the natural wind.

In the first section of the paper non-dimensional parameters are introduced which may be used to define the properties of a model required to produce dynamic similarity with the full-scale structure. A separate section discusses the requirements for matching the airstream in the wind tunnel to that of the natural wind prevailing at the site of the structure. The paper is completed by a discussion of the errors introduced into force measurements on wind-tunnel models by an interference effect known as wake blockage. This effect is of particular significance for tests on building models owing to the wide wakes generated by such models. A correction procedure is introduced which permits the use of larger wind-tunnel models than would otherwise be possible without serious measurement errors being incurred.

2. LIST OF SYMBOLS

The suffices m and f are used to denote model and full-scale values respectively.

A   aspect ratio,
suffices 1 and 2 refer to the values for the cross wind force-drag and lift-drag planes respectively

B	typical external linear dimension
C	area of wind-tunnel working section
C_L, C_D, C_C	lift, drag and cross wind force coefficients respectively
C_p, C_p, C_F, C_M	pressure, force and moment coefficients
c	model chord
D	diameter
E	modulus of elasticity
F	total wind force
g	gravitational acceleration
h	linear dimension used to define a moment coefficient
I	moment of inertia per unit span, suffices θ and z refer to torsional and bending modes respectively
K_s	structural damping coefficient
k	numerical constant in equation 5.3
M	overturning moment due to wind
m	mass per unit length
N	frequency
p, p_w	local surface pressure, suffix w refers to the windward face
Re	Reynolds number $\left(= \dfrac{\rho VB}{\mu}\right)$
S	area used to define the force and moment coefficients, suffices m and g relate respectively to the model and to any obstruction in the airstream not directly included in the model measurements
s	model span

t	wall thickness
V	wind speed, suffices z and z_o refer to wind speeds at these heights
z	height above ground, suffix o is used to denote a reference height
$1/\alpha$	exponent of wind-speed profile for long period means, and specifically for the mean-hourly speed
δ	logarithmic decrement, defined as the natural logarithm of the ratio of successive amplitudes of oscillation, suffices a and s refer to the aerodynamic and structural components respectively
μ	dynamic viscosity of air
ρ	density of air
σ	structural density

Other symbols used are defined where they appear in the text.

3. THE SIMILARITY REQUIREMENTS FOR AEROELASTIC SCALING

3.1 *The Non-Dimensional Parameters and Their Application to Model Scaling*

The laws which relate the results of model experiments to the behaviour of the full-scale structure are most easily and concisely determined by the method of dimensional analysis. Provided that all physical quantities involved in the phenomenon under consideration are incorporated into the analysis, the method provides the functional relationship between the non-dimensional values of these quantities and yields the similarity laws permitting full-scale predictions to be made from model-test results.

It can be postulated that the behaviour of a structure under steady uniform wind conditions will depend on the aerodynamic shape and on the following eight physical quantities:-

$$E, \rho, V, B, \mu, g, \sigma \text{ and } \delta_s$$

where each is defined in the list of symbols.

From these eight quantities the five non-dimensional parameters given in Table I can be derived. These parameters may be regarded qualitatively as ratios of forces or energies and the appropriate ratio for each parameter is shown in the Table.

Table I

The Non-Dimensional Parameters

Non-Dimensional Parameter	Symbolic Definition	Force or Energy Ratio
1. Logarithmic Decrement	δ_s	$\dfrac{\text{Energy dissipated per cycle}}{\text{Total energy of Oscillation}}$
2. Elasticity	$E/\rho V^2$ *	$\dfrac{\text{Elastic forces of the Structure}}{\text{Inertia forces of the air}}$
3. Density Ratio	σ/ρ *	$\dfrac{\text{Inertia forces of the structure}}{\text{Inertia forces of the air}}$
4. Gravitational	gB/V^2	$\dfrac{\text{Gravitational forces on the structure}}{\text{Inertia forces of the air}}$
5. Reynolds number	$\rho VB/\mu$	$\dfrac{\text{Inertia forces of the air}}{\text{Viscous forces of the air}}$

* either of these parameters may be replaced by $\dfrac{V}{NB}$

3.2 *The Design of Wind-Tunnel Models*

In practice it is rarely possible to satisfy all the non-dimensional parameters listed in Table I with a model of reduced linear scale tested in an atmospheric wind tunnel, both because the conditions imposed by the individual parameters are incompatible and because in general no materials for model construction are available with the physical properties demanded by the similarity requirements. For this reason it is usually necessary to relax the similarity requirements to some extent and to introduce compromise procedures which, while introducing uncertainties into the interpretation of the model results, still enable relevant data to be obtained.

The selection of the similarity parameters from those listed in Table I which need to be satisfied in any particular instance depends upon the nature of the problem to be investigated. A diagram showing the types of wind-tunnel model used for investigation of various wind loading problems is given in Fig. 1. In the following sections the similarity requirements for the three main types of model listed in Fig. 1 are considered in detail.

3.3 *Static Rigid Models*

Static rigid models are used for the determination of surface pressures and overall steady forces and moments experienced by a building or structure. Full-scale values of the Reynolds number cannot be achieved on reduced scale models tested in atmospheric wind tunnels and for each particular investigation it is therefore necessary to consider the extent to which the force and pressure coefficients (defined below) vary with Reynolds number. It has been shown by tests carried out over a wide range of Reynolds number that the airflow pattern around a sharp-edged body is substantially independent of this parameter because the positions of the flow separations are fixed by the sharp edges. Whether this independence extends to the extreme values of Reynolds number obtained on full-scale buildings is, as yet, unconfirmed and awaits the results of attempts to correlate model results with measurements on the full-scale. Nevertheless for many building shapes it is considered reasonable to assume an independence of the Reynolds number and to carry out tests in a wind tunnel on a model linearly scaled to be as large as possible without introducing unduly large interference effects due to constriction of the airflow by the wind-tunnel walls (see section 5).

For bodies of rounded shape, flow separation is not fixed by the presence of a sharp edge and the flow pattern may vary considerably with Reynolds number so that coefficients measured on a model at low Reynolds numbers may be significantly different from those experienced by the full-scale building. Attempts have been made to reproduce on models the flow conditions at high Reynolds numbers either by roughening the windward surface of the model or by using boundary-layer trip wires, so that the flow separations are induced to occur at similar positions on the model as on the full-scale, but the extent to which these devices are successful is in some doubt.

Measurements made on rigid models have their most general application when expressed in the form of non-dimensional coefficients. These coefficients, which may be used to determine the wind loading at any wind speed on structures of similar external shape, are generally defined as follows:-

(i) Pressure coefficient $C_p = \dfrac{p}{\frac{1}{2}\rho V^2}$

(ii) Force coefficient $C_F = \dfrac{F}{\frac{1}{2}\rho V^2 S}$

(iii) Moment coefficient $C_M = \dfrac{M}{\frac{1}{2}\rho V^2 S h}$

where S and h refer to the frontal area and height of the building respectively.

3.4 *Dynamic Models*

Strictly, models used to examine the oscillatory behaviour of a structure in wind, should reproduce the correct values of all the structural parameters 1-4 in Table I together with the correct aerodynamic shape and value of the Reynolds Number. This last parameter will be subject to the same problems discussed in the section on static rigid models. For observations of static deflection and stress due to wind only the elastic and gravitational parameters need to be considered (see section 3.5) but for for dynamic studies both the inertial and the damping parameters have to be taken into account.

Two methods of test have been used to study the dynamic behaviour of structures in wind. The first involves the use of a model of the complete structure, termed a <u>full model</u>, in which the similarity parameters are satisfied either by a replica reproduction of full-scale in suitable materials or by a different type of construction chosen to yield the required mass and stiffness distribution. The alternative method uses spring-mounted rigid models of typical lengths of the structure which are termed <u>sectional models</u>. The use of such models is applicable to tests on structures such as long-span suspension bridges and tall lattice masts where the reduction in the linear scale required to accommodate the complete model in the wind tunnel would so reduce the dimensions of the individual components of the model that either its construction would be impracticable or the results would be subject to unacceptable Reynolds number effects.

In the following paragraphs the methods of satisfying the similarity requirements on models will be discussed for two types of structure.

3.4.1 *Full Models of Suspension Bridges*

The full model must have an external shape similar to that of the full-scale although true reproduction of the internal structural details is not necessary provided the inertial and elastic stiffness requirements can be maintained. Model and full-scale air densities are the same for tests in an atmospheric wind tunnel. The effect of this is twofold. Firstly the correct representation of Reynolds number is not possible and the remarks made in the previous section concerning the justification for the neglect of this parameter are applicable. Secondly, for consistency of the density ratio, the structural densities at all corresponding positions must be the same on the model as on the full-scale.

Velocity scaling is prescribed by the gravitational parameter as,

$$\left[\frac{gB}{V^2}\right]_m = \left[\frac{gB}{V^2}\right]_f \qquad \ldots (3.1)$$

and since the gravitational acceleration must be the same for both systems,

$$\frac{V_m}{V_f} = \sqrt{\frac{B_m}{B_f}} \qquad \ldots (3.2)$$

To be compatible with this prescribed velocity scale the required elastic properties can only be achieved on a replica model by the use of materials for model construction with moduli of elasticity given by

$$\left[\frac{E}{\rho V^2}\right]_m = \left[\frac{E}{\rho V^2}\right]_f \qquad \ldots (3.3)$$

so that

$$\frac{E_m}{E_f} = \frac{B_m}{B_f} \qquad \ldots (3.4)$$

Suitable materials are not usually available and compromise procedures have to be adopted which yield an equivalent stiffness effect. These usually involve either an overall reduction in the cross-sectional areas of the elastic members; (the necessary restoration of the external shape and mass distribution being achieved by the addition of discrete fairings and masses), or by the adoption of an entirely different method of construction. In the construction of their full model Frazer and Scruton[1] adopted the former method for the model suspension cable, the stiffness properties of which were obtained by a steel wire of area reduced by the ratio B_M/B_f from that of the dimensionally-scaled model cable, and the shape and mass distribution were restored by the point attachment of brass cylinders along the length of the cable. The latter principle was adopted for the suspended structure; the road decks and stiffening trusses being made up of short rigid segments of the correct external shape and interconnected by small steel springs to provide the correct overall elastic stiffnesses.

The introduction of a representative amount of structural damping into the model tests presents certain fundamental difficulties since it is neither possible to predict a value of the full-scale damping nor to adopt a model construction which automatically achieves the correct value of this parameter. The uncertainties surrounding the structural damping cannot be adequately resolved and, in order that the stability predictions may err on the safe side, tests on models need to be performed with values of damping well below the minimum value likely to occur on the full-scale bridge.

3.4.2. *Sectional Models of Suspension Bridges*

In this method of test the conditions for complete dynamic similarity are obviously not observed. It is assumed that the aerodynamic forces which cause the various modes of oscillation of the full-scale structure are the same as those producing the rigid-body oscillations of the sectional model. The validity of this assumption has been demonstrated experimentally by Frazer and Scruton[1]. They show that for sectional models, the similarity requirements are that the external shape and the following non-dimensional parameters (the equivalent expressions for parameters 1, 2 and 3 in Table I) shall be the same for model as for full-scale:-

(a) Damping — $\delta_{\theta s}$, δ_{zs}

(b) Reduced Velocity — $\dfrac{V}{N_\theta B}$, $\dfrac{V}{N_z B}$

(c) Inertia — $\dfrac{I_\theta{}^*}{\rho B^4}$, $\dfrac{I_z{}^*}{\rho B^2}$

where the suffices θ and z refer to torsional and bending modes of oscillation.

---

* The values of I_θ and I_z are the values per unit span of the bridge and include the contributions of the cables. For both vertical bending and torsional oscillation the cables are assumed to move vertically.

Rigid sectional models are usually of the light-weight construction required to reproduce the inertial characteristics of the full-scale bridge as specified by parameter (c). They are mounted across the wind tunnel on a spring suspension so that two-dimensional flow conditions are maintained over the entire length of the model, and oscillation can take place in vertical or pitching motions, representing respectively the vertical bending and torsional oscillation of the full-scale bridge. The stiffnesses and arrangement of the supporting springs are chosen to provide the appropriate frequency ratio N_θ/N_z with values of N_θ and N_z giving a convenient wind-speed scale. The suspension is designed to have as low a value of damping as possible, well below the expected full-scale value, so that various values of controlled external damping may be applied to the model.

In some cases it may not be possible to construct a model light enough to satisfy the inertial requirements and at the same time be rigid enough for the purposes of the tests. In such instances, Scruton[2] has shown that it is permissible to combine the parameters (a) and (c) above to form a single parameter

$$\frac{I_\theta \delta_{\theta s}}{\rho B^4} \quad , \quad \frac{I_z \delta_{zs}}{\rho B^2}$$

provided that the frequency of oscillation does not vary with wind speed, thus indicating that the in-phase component of the aerodynamic force is negligibly small.

The alternative scaling procedure may be used for the different modes of oscillation provided that the modes occur in isolation. For the study of oscillatory behaviour involving coupling between different modes, or if the results of the tests are to be used for amplitude predicition, then proper inertial scaling is required.

3.4.3 *Full Models of Slender Towers and Stacks*

The discussion of the similarity requirements given in section 3.4.1 is equally applicable to the problem of reproducing the oscillatory behaviour of towers and stacks on models of reduced linear scale. For towers and stacks the influence of the gravitational parameter is usually insignificant and the similarity parameters to be satisfied on the model may be expressed as follows:-

(1) δ_s

(2) $\dfrac{Et}{\rho V^2 B}$ (equivalent to $\dfrac{E}{\rho V^2}$ in Table I)

(3) $\dfrac{m}{\rho B^2}$ (equivalent to $\dfrac{\sigma}{\rho}$ in Table I)

(4) $\dfrac{\rho V B}{\mu}$

The use of the alternative form of the elasticity parameter is applicable to thin-walled structures and to lattice towers where the distance of the stress bearing members from the neutral axis is large compared with the thickness t. It is useful in that it enables convenient materials of construction and wind-speed scales to be adopted since the requisite stiffness effect Et may be obtained on the model by adjustment of the thickness scale.

A model designed to satisfy the stiffness requirement will not, in general, satisfy the inertial requirement $\dfrac{m}{\rho B^2}$ (where m is the mass per unit length). Provided that the inertial value has not already been exceeded in achieving the required stiffness then it is possible to increase the inertia by the addition of discrete masses at intervals along the height of the model. The location and value of each of these added masses are calculated to bring the energy of oscillation of the significant modes up to the required value. As for the bridge models, independent scaling of the inertia and damping parameters is not essential if there is no frequency change with wind speed so that (1) and (3) above may be combined to form a single parameter $\dfrac{m \delta_s}{\rho B^2}$. If inertial scaling is not attempted then the speed scale is no longer prescribed by the elastic scaling parameter but only by the reduced velocity parameter $\dfrac{V}{NB}$ so that

$$\dfrac{V_m}{V_f} = \dfrac{N_m}{N_f} \cdot \dfrac{B_m}{B_f} \qquad \ldots (3.5)$$

3.4.4 *Sectional Models of Slender Towers and Stacks*

The similarity requirements for the sectional-model testing of towers and stacks are the same as those discussed in the section 3.4.2 for suspension bridges. However, the objectives of the test are usually somewhat different. The suspended structure of a bridge is uniform across the whole span so that tests can be limited to a small range of wind direction with respect to the spanwise axis, and structurally acceptable modifications of the shape to improve the aerodynamic stability are readily made. Tests on bridges are therefore usually directed towards ensuring stability by modification to the aerodynamic shape. The same consideration does not apply to slender towers and stacks, and tests of these structures are usually directed towards the measurement of the aerodynamic excitation so that a stability prediction can be made by taking into account the anticipated amount of structural damping. To obtain the total aerodynamic excitation experienced by a full-scale tower it is necessary to measure the excitation on a number of sectional models each representing a different typical section of the tower and to integrate, using a strip theory, this excitation over the full height. For this purpose calculated frequencies and modal shapes are used.

3.5 *Static Aeroelastic Models*

The distortions and stresses in the structural members of a proposed building or structure due to wind may be determined by wind-tunnel tests of an aeroelastic model, provided that the model is built as a replica of the full-scale with all the relevant structural detail reproduced. The similarity requirements for static aeroelastic models differ from those for the dynamic aeroelastic types discussed in sections 3.4.1 and 3.4.3 in that, for the static model, the density and gravitational effects have only a minor influence. Consequently parameters 3 and 4 in Table I may be disregarded. Dynamic models which satisfy the condition of replica aeroelastic construction are, of course, equally suitable for distortion and stress measurements.

A dynamic aeroelastic model of a tall building block has been built in perspex at the NPL[3] and used for a wind-tunnel study of the distortion and stresses in structural members due to wind. The dynamic scaling of the model also enabled observations of aerodynamic stability to be included as part of the investigation.

4. THE REPRESENTATION OF NATURAL WIND CHARACTERISTICS IN THE WIND TUNNEL

4.1 *The Vertical Gradient of Velocity and Its Effect on Airflow around Structures*

Apart from thermal effects the variation in speed of the natural wind with height, known generally as the vertical wind gradient, is dependent on the relative roughness of the local terrain. Measurements of the gradient in different localities have shown that it is reasonable to represent the variation of speed (V_z) with height (z) as a simple power law variation of the form

$$V_z = V_{z_o} \left(\frac{z}{z_o}\right)^{1/\alpha} \qquad \ldots (4.1)$$

where V_{z_o} is the speed at a reference height z_o. For mean-hourly wind speeds Davenport[4] has suggested values of $1/\alpha$ ranging from $1/7.5$ for open grassland to $1/4$ for heavily built-up areas; these exponents are decreased for wind speeds averaged over a shorter period.

Many of the shape and pressure coefficients used to specify wind loading on buildings and structures have been obtained from tests in wind tunnels with wind streams uniform except for the natural boundary layer of the wind tunnel. Attempts to deduce the wind loading on a building immersed in an airstream with a vertical velocity gradient from results obtained in a uniform stream, have been based on the simple assumption that the gradient produces no overall change in the flow pattern, so that the wind load per unit length at any height (z), is proportional to the square of the speed of the approaching wind at that height.

i.e. Force per unit length $= \frac{1}{2}\rho V_z^2 C_F \qquad \ldots (4.2)$

where C_F is the overall shape factor for the building which has been determined from tests in a uniform wind.

Wind-tunnel tests have shown that this procedure is unsound since the variation of the wind load with height is felt only by the windward face, while the suction on the leeward face is approximately constant in magnitude over the entire height. Baines's[5] investigation on a tall building immersed in an airstream with a vertical velocity gradient showed that the measured pressure distribution on the windward face was in reasonable agreement with the distribution predicted by

$$p_w = \tfrac{1}{2}\rho V_z^2 C_{p_w} \qquad \ldots (4.3)$$

where C_{p_w} is the pressure coefficient on the windward face determined from tests in a uniform wind. In addition Baines[5] has suggested that pressures on the leeward face in a wind gradient can be taken as some 60% of the values recorded in tests in a uniform airstream where the reference velocity is taken as that corresponding to the top of the building.

Jensen[6], on the other hand, is of the opinion that wind-tunnel tests on low buildings under uniform wind conditions are misleading and that a correct representation of the 'Model Law' is essential.

4.2 *The Production of a Velocity Gradient in the Wind Tunnel*

The methods so far used in wind-tunnel investigations rely, in the main, on the introduction of a graduated resistance into an initially uniform airstream so that for some considerable distance downstream a steady velocity gradient is maintained with a uniform static pressure throughout. In some cases a simple empirical approach has yielded good results. O'Neill[7] produced the required graduation in resistance by spacing rods of different diameters in a plane normal to the airstream. Baines[5] used a similar empirical approach by introducing a curved wire-mesh screen some distance ahead of a model to produce a gradient having an exponent $1/\alpha = 1/6$. Jensen's Model Law[6] suggests a somewhat different approach which involves representing the correct scale of ground roughness for a considerable distance upstream of the model so that the wind gradient is formed in similar manner to that of the natural wind. The disadvantage of this method appears to be that, in order to reproduce the ground roughness effect for a reasonable full-scale distance upstream of the model in the limited length of working section available in most wind tunnels, very small-scale models have to be used. This restriction does not apply to the methods used by O'Neill[7] and Baines[5] although their purely empirical approach has the disadvantage that considerable time is usually expended in obtaining the required velocity profile. The analytic approach of Owen and Zienkiewicz[8] provides a useful method for calculating an arrangement for a grid of rods to produce a uniform shear flow whilst Elder[9] has calculated the effect of using gauzes of non-uniform resistance to produce various velocity profiles. Both these approaches show good agreement between theory and experiment and it seems likely that one or both of these methods will form a suitable basis for representing natural wind gradients in the wind tunnel.

4.3 *Atmospheric Gusting*

The gust spectrum present in the natural wind arises from the diverse eddy systems produced by the viscous interaction between the atmosphere and the ground surface. Until recently the effects of gusting winds have been taken into account by specifying a design wind speed equal to that of the gust speed and computing the wind loading as for a steady wind of this speed. The gust speed referred to here is a maximum mean speed over a short period of time and, of course, varies with the duration chosen. The gust duration and hence the design wind speed is usually related to the time taken for the gust to develop over the whole structure and to the response characteristics of the structure itself.

Davenport[10] had adopted a more rigorous approach and using the concept of a stationary time series has determined the response of a simple structure to a turbulent gusty wind in terms of the resultant stresses and deflections.

4.4 *Possibilities of Producing Gust Effects in Wind-Tunnel Testing*

To make use of the turbulence levels available in existing wind tunnels, models would have to be impracticably small. The methods described in the previous section for the production of a velocity gradient by means of grids or screens do not in general reproduce the right scale of turbulence. On the other hand the method of roughening the surface of the wind-tunnel floor in the manner suggested by Jensen[6] would appear to reproduce both the gradient and gust effects at the same time. In fact the 'Model Law' approach appears to be the only method of reproducing a gust spectrum representative of the natural wind although in some cases the use of wire screens or perforated plates may be useful in producing turbulence of a particular intensity.

5. THE EFFECTS OF WIND-TUNNEL-WALL INTERFERENCE ON MODEL RESULTS

5.1 *The Interference Effect*

Full-scale wind loads based directly on uncorrected wind-tunnel measurements on a model situated within an enclosed working section will tend to be overestimated because extraneous suction effects are introduced into the wake (termed "Wake blockage") due to the proximity of the tunnel walls. The extent to which this overestimation occurs is directly related to the proportion of the tunnel cross-sectional area occupied by the model. It is further dependent on whether the flow over the model is attached or separated. In order to limit the errors introduced into wind-tunnel force and pressure measurements to an acceptable amount therefore, it is desirable to limit the size of model so that less than 5 per cent of the area of the working section of the wind tunnel is occupied by the model. In cases where the use of larger models is unavoidable, it is essential to apply corrections for the effect of wake blockage. Investigation of this effect has shown that simple empirical corrections may be applied to force and pressure measurements although these would not be expected to provide sufficient accuracy for models occupying more than 10 per cent of the tunnel working section area.

The interference effect is associated with the flow downstream of the model, where the reduced value of the stream-wise velocity in the wake region requires an increase in speed of the air outside the wake region to maintain continuity of mass flow across the section of the tunnel. This increase in speed is balanced by a decrease in static pressure of the air outside the wake since the stagnation pressure* of this air is unaffected by the presence of the model. Consequently, since the static pressure within the wake is governed by that of the steady airstream immediately adjacent to the boundary of the wake, the static pressure in the lee of the model (termed the base pressure) tends to be less than it would be if the airstream were unconfined as in the full-scale case. For this reason wind forces on models, in particular the drag, tend to be overestimated in wind-tunnel tests.

The magnitude of the change in base pressure is dependent on the flow conditions prevailing on the model. For a stream-lined shape where the flow remains attached over the entire model surface the wake is very narrow and the correction required is much smaller than that for a model of similar cross-sectional area where the flow is completely separated and the wake is as wide if not wider than the model itself. The wake blockage correction developed for stream-lined shapes has proved inadequate for tests on bluff building shapes where flows are mainly of the completely separated type. In the following section the correction equation used in NPL tests on bluff building shapes is presented and details of its application are discussed.

5.2 *The Correction for Wake Blockage*

A complete expression for wake blockage suggested by Maskell[11] is quoted below. In the quoted form it is applicable to a model on which separation of the flow occurs towards the rear of the model so that both components for attached and separated flows are included in the equation:-

$$\frac{C_{D_m}}{C_{D_F}} = 1 + 0.5\left[C_{D_o}\frac{S_m}{C} + C_{D_g}\frac{S_g}{C}\right] - \frac{1}{C_p}\left[C_{D_m} - C_{D_o} - C_{D_I}\right]\frac{S_m}{C} \quad \ldots (5.1)$$

where C_{D_m} is the measured value of C_D based on an area S_m

C_{D_F} is the true value of C_D for the free-stream condition

C_{D_o} is the minimum value C_D for which $C_L = C_C = 0$ for completely attached flow

C_{D_I} is the lift and cross wind force contribution to C_D

---

* The pressure developed by an airstream when it is brought to rest.

C_{D_g} is the value of C_D, based on an area S_g, for any obstruction in the airstream which is not directly included in the model measurements

C_p is the base pressure coefficient in the separated wake behind a bluff body. For a flat plate normal to the airstream $C_p = -0.4$ in the aspect ratio range 1 to 5 so that for a general case $-\frac{1}{C_p} = 2.5$.

A detailed discussion of this equation will not be given in this paper but will appear in a paper to be published by Maskell[11]. For the present purposes the flow around bluff building shapes is regarded as completely separated and so that only the part of Maskell's equation which deals with this type of flow* will be considered. The correction for completely separated flow may be written therefore as:-

$$\frac{C_{D_m}}{C_{D_F}} = 1 - \frac{1}{C_p}\left[C_{D_m} - C_{D_I}\right]\frac{S_m}{C} \qquad \ldots (5.2)$$

Here C_{D_o} and C_{D_g} have been omitted from equation 5.1 because C_{D_o} no longer represents a real quantity and it is assumed that the model is supported internally so that there is no obstruction to the flow other than the model itself. Although equations 5.1 and 5.2 derive a correction based primarily on drag measurement it is considered that the same percentage correction is applicable also to the other measured forces and moments.

The simplified correction equation 5.2 is considered to be adequate for most tests on building shapes so that the only information additional to the measured drag, required to apply the correction is a value for the coefficient C_{D_I}. An estimate of this so-called induced drag coefficient may be obtained by considering the induced drag equation derived for aerofoil theory:-

$$C_{D_I} = \frac{k}{\pi}\left(\frac{C_c^2}{A_1} + \frac{C_L^2}{A_2}\right) \qquad \ldots (5.3)$$

---

* Particular cases, such as reflector dishes, may experience attached flow for certain attitudes of the dish, in which case the complete correction given by equation 5.1 is required.

where A_1 is the aspect ratio appropriate to forces in the cross wind force-drag plane and A_2 to those in the lift-drag plane. The value of k will, in general, be determinate only in cases of attached flow where k is the slope of the C_{D_m} versus $\frac{1}{\pi}\left(\frac{C_c^2}{A_1}+\frac{C_L^2}{A_2}\right)$ curve at the minimum value of C_{D_m}. The extent to which this method may be applied to shapes with completely separated flows is limited by the non-linearity of the C_{D_m} versus $\frac{1}{\pi}\left(\frac{C_c^2}{A_1}+\frac{C_L^2}{A_2}\right)$ curve; in general it has not been found possible to ascertain the value of k by direct measurement. However for most structural shapes C_{D_I} is small and it is considered that a correction based on k = 1 provides sufficient accuracy.

Equation 5.2 is intended to be applied to tests where the wind upstream of the model has a uniform velocity profile. Measurements made in non-uniform velocity profiles have been corrected using the same equation although the effect of a velocity profile on wake blockage is uncertain.

ACKNOWLEDGEMENT

The author wishes to acknowledge the helpful advice given by Mr. C. Scruton and by Mr. R. W. F. Gould, both of the Aerodynamics Division, National Physical Laboratory.

The paper is published by permission of the Director of the National Physical Laboratory.

REFERENCES

1. FRASER, R. A. and SCRUTON, C.   A summarised account of the Severn Bridge Aerodynamic Investigation.
   NPL/Aero Report 222, HMSO, 1952.

2. SCRUTON, C.   An experimental investigation of the aerodynamic stability of suspension bridges.   *Prel. Pub. for the 3rd Congress Inter. Assoc. Bridge and Struct. Engrg., 1948.*

3. SCRUTON, C., WHITBREAD, R. E. and CHARLTON, T. M.   An aerodynamic investigation for the 437 ft Tower Block proposed for the Albert Embankment, Vauxhall, London.   *NPL Aero Report 1032, 1962.*

4. DAVENPORT, A. G. Rationale for determining design wind velocities. *Proc. Amer. Soc. civ. Engrs. J. struct. Div.*, 1960, 86, ST5.

5. BAINES, W. D. Effect of velocity distribution on wind loads on tall buildings. *University of Toronto*, TP6203, June 1962.

6. JENSEN, M. The model-law for phenomena in natural wind. *Ingeniøren-International Ed.*, 1958, 2, 4, 121.

7. O'NEILL, P. G. G. Experiments to simulate a natural wind gradient in a compressed air tunnel. *Unpub. report, NPL/Aero/313*, 1956.

8. OWEN, P. R. and ZIENKEIWICZ, H. K. The production of uniform shear flow in a wind tunnel. *J. fluid. Mechs.*, 1957, 2, 6.

9. ELDER, J. W. Steady flow through non-uniform gauzes of arbitrary shape. *J. fluid Mechs.*, 1959, 5, 3.

10. DAVENPORT, A. G. The application of statistical concepts to the wind loading of structures. *Proc. Instn. civ. Engrs.*, 1961, 19, 449.

11. MASKELL, E. C. A theory of the blockage effects on bluff bodies and stalled wings in a closed wind tunnel. *R.A.E. Report No. Aero 2685*, November 1963.

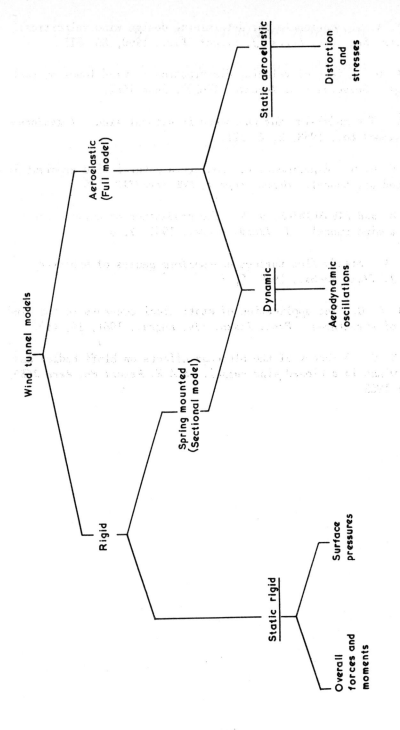

Fig.1. Wind-tunnel models and their application to wind loading problems

DISCUSSION ON PAPERS 5, 22, AND 21

MR. CLARK mentioned the failure, through what was later attributed to wind-induced oscillations, of a form of aluminium railing in which 5/8 inch diameter bars were carried vertically between 2 inch diameter tubes of wall thickness 1/4 inch. The bars vibrated at 240 cycles/sec in winds of 14 knots, and their failure was the result of the consequent fatigue. The remedy was to cut them through and to put on a sleeve. He enquired whether this type of problem could be simulated in a wind tunnel. Even if these railings could have been so tested, he knew of similar happenings on string bow bridges which could not be put into a wind tunnel.

PROFESSOR OWEN referred to the simulation of wind shear in model tests and to the simulation of the buffeting to which a structure is exposed because of atmospheric turbulence. The two must be treated as separate issues. In steady flow it appeared incontrovertible that wind shear becomes important if the shear parameter is of the order of unity. The shear parameter is the difference between the velocity at the top of the structure and that at some point where it is zero, divided by the mean. He had not found it difficult to produce velocity gradients, with a shear parameter of order unity, by means of grids. It is not essential to produce wind profiles near the ground very precisely, but he wished to make the plea that many structures should be tested in a linear velocity gradient over the whole of the wind tunnel. It is possible to have a number of grids, or curved or graded screens, which could be put into the wind tunnel at will, to give a series of standard velocity gradients. The shear parameter is of tremendous importance and is difficult enough to deal with. The unsteady flow problems, however, are much more so. In consideration of the turbulence it is necessary to recognise that turbulence depends not only on Reynolds number but also on the Richardson number (the Richardson number is the measure of the work done by the buoyancy force in the atmosphere to that done by turbulence). Both parameters have to be reproduced correctly if the spectrum of turbulence is to be correctly reproduced, and to do this the powers required to drive the wind tunnel are comparable with those needed to drive the wind round the earth. For dynamic similarity further complications arise because of the additional requirement to satisfy the other parameters, e.g. the inertial and the elastic parameters. It would seem preferable therefore not to attempt the impossible but to limit the problem to examining the response of a structure to certain discreet frequencies. It may then be possible to calculate the response of a structure to the full spectrum of atmospheric turbulence from the kind of data which has been presented by Professor Davenport.

MR. GOULD said that he was pleased to note from MM Colin and d'Havé's Paper that the results of wind-tunnel tests with velocity gradients represented appeared to apply to real buildings. With regard to the upstream effects on buildings, Lighthill calculated the boundary layer separation position for an infinitely long wall, and found that the turbulent boundary separated upstream at a distance 3.9 times the wall height, and this distance was only reduced to about one-half with the wall reduced to a length of three and a half times its height. This result is significant in relation to wind-tunnel testing and implies that a very long working section is not necessary. The vortex which is produced in front of a building results from separation. It is not possible to terminate a vortex in space and so this vortex must curve round the building in a horseshoe fashion. Professor Baines ascribed the downflow on the windward surface of the building to difference in static pressure between the top and the bottom. However, the vortex will induce downward flow and so it is a question of which comes first, the vortex or the static pressure gradient? If the static pressure gradient is caused by the curvature of the streamline it is not so important to reproduce the velocity profile as to have a boundary layer which will separate at the correct position.

It is surprising how wind-tunnel blockage can affect the static pressures. A quite small radar reflector of 18 inches diameter in a 7 ft x 7 ft wind tunnel can produce a base pressure coefficient of -0.8, which in a free stream would be only -0.4. Blockage corrections in pressure measurements are therefore very important.

DR. PRIS reiterated that there was a marked difference in viewpoint between those who demanded tests in a velocity gradient and those who thought that this was not necessary. One of his objections to the velocity gradient approach was that there is an infinite number of widely divergent gradients as shown by the values of the exponent in the power law suggested by different writers. With such widely different gradients it could not be claimed that more precision is obtained by reproducing a velocity gradient. It is to be doubted whether the vortex in front of a building shown by the flow photographs in MM Colin and d'Havé's Paper really exists on the full-scale building, and hence there is no similarity in this case between the wind-tunnel model and the real building.

DR. PRIS (written contribution)

1. Les communications qui viennent d'être présentées, montrent qu'il existe deux écoles nettement caractérisées concernant la préparation des essais sur maquettes de bâtiments au laboratoire aérodynamique.

Pour l'une (communications 15 and 22), des mesures précises ne peuvent être obtenues en laboratoire, et il ne peut y avoir similitude complète entre les essais et la vraie grandeur si le gradient des vitesses du vent n'est pas reproduit à l'échelle en soufflerie; pour l'autre (communications 6, 21 et 5), cette représentation n'est pas nécessaire et les essais ordinaires avec vitesse uniforme du courant en soufflerie, permettent de déterminer correctement les pressions sur bâtiments en vraie grandeur.

Une étude critique des deux méthodes est donc nécessaire. La similitude étant réalisée en ce qui regarde les maquettes elles-mêmes, les points suivants doivent seuls être considérés:

- réalisation d'un gradient déterminé des vitesses en soufflerie;

- utilisation des valeurs numériques obtenues par l'une et l'autre méthodes pour le calcul des pressions en vraie grandeur.

2. Une formule simple est souvent adoptée pour représenter l'évolution des vitesses en fonction de la hauteur H au dessus du sol;

$$\frac{V_H}{V_o} = \left(\frac{H}{H_o}\right)^n$$

Suivant les auteurs, l'exposant n'est compris entre $\frac{1}{2,5}$ et $\frac{1}{12}$, ce qui traduit l'influence des nombreuses variables.

Comme il ne peut être question d'effectuer avant toute construction un relevé du gradient local, on voit que la remarque précédente retire beaucoup de la précision cherchée en utilisant un montage avec gradient en soufflerie.

3. Supposons que ce montage puisse néanmoins être effectué avec précision à l'échelle de la maquette, en réduisant les hauteurs H au $1/100^e$, $1/200^e$ et même au $1/500^e$, puisque des maquettes de 5 à 7 cm sont parfois utilisées. Entre les gradients $G_1 = \left(\frac{\Delta V_H}{\Delta H}\right)_1$ vraie grandeur et $G_2 = \left(\frac{\Delta V_H}{\Delta H}\right)_2$ soufflerie, on aura la relation:

$$G_2 = 100 \cdots 200 \cdots 500 \times G_1$$

On sait que les gradients du vent G_1 ont toujours une valeur faible par suite du mélange des couches d'air inférieures. Au contraire, la théorie de Prandtl, ainsi que les mesures déjà faites, montrent que, dans une couche limite à la surface d'une plaque ou d'une paroi représentant le sol, le gradient tend vers une valeur infinie (com. 4, fig. 8 et com. 5, fig. 7 and 8). Il est alors prouvé (comm. 5) qu'il existe un point de décollement en amont de la face avant des maquettes, suivi d'un tourbillon à axe horizontal bien visible sur les fig. 13 and 14 de la communication 22.

Les pressions mesurées sur les faces des maquettes, surtout des maquettes allongées, dépendront essentiellement de l'intensité de ce tourbillon lequel est inexistant ou d'intensité négligeable en présence d'un bâtiment en vraie grandeur soumis à l'action du vent à faible gradient.
4. Si, au contraire, la vitesse V_s en soufflerie est uniforme, les pressions en vraie grandeur seront calculées comme le propose le pr. Baines (Comm. 6), en opérant par tranches pour les faces situées en amont de la ligne de décollement. Soient H et h les hauteurs correspondantes:

p_H vraie grandeur = $c_s \cdot q_H$, c_s coefficient de soufflerie.

On aurait: p_h en soufflerie = $c_s \cdot q_s$, avec $q_s > q_H$ -

Pour les surfaces situées en arrière de la ligne de décollement, il y aurait peut-être lieu d'appliquer aux valeurs de c un coefficient de correction, mais on doit remarquer que au cours des essais signalés par le Pr. Baines, le tourbillon de façade n'était pas éliminé.
5. Il en est de même avec les essais signalés par MM. Niels Franck (comm. 15) et Colin et d'Havé (comm. 22).
6. En conclusion, tout effet de couche limite doit être éliminé au cours des essais en soufflerie. Ce n'est qu'alors que des comparaisons valables pourront être effectuées entre pression sur maquette et en vraie grandeur, lorsque ces dernières pourront faire l'objet de mesures précises.

H. COLIN (in reply). It appeared from our tests that similar pressure distributions were obtained irrespective of whether or not the velocity gradient covered the whole building, although there may be some displacement in the absolute values of the pressures. His tests had been carried out to provide data for revising the existing code of practice and it was necessary to simplify and to standardise on the building shapes to be tested. Tests carried out in a thin boundary layer probably lead to wind loading estimates which are on the safe side, and that was why they chose the method of test described in the Paper. It would be useful if all workers in the field decided on some definite testing method which would yield results sufficiently accurate for incorporation in the codes of practice.

MR. WHITBREAD (in reply). At the National Physical Laboratory we have been most concerned with either very tall structures or long span bridges for which, it is believed, the effects of velocity gradient are not very significant.

SESSION 4

Chairman:

    Dr. F. M. Lea
    (Director, Building Research Station, U.K.)

PAPER 19

LES REGLES FRANÇAISES 1963 DEFINISSANT LES EFFETS DU VENT SUR LES CONSTRUCTIONS

by

N. ESQUILLAN
(Président de la Commission Technique pour l'étude
des effets de la neige et due vent sur les constructions. Président de la Commission de rédaction
des Règles Neige-Vent)

LES REGLES FRANÇAISES 1963 DEFINISSANT
LES EFFETS DU VENT SUR LES CONSTRUCTIONS

by

N. ESQUILLAN
(Président de la Commission Technique pour l'Étude des Effets
de la Neige et du Vent sur les Constructions Président de la
Commission de Rédaction des Règles Neige-Vent)

EN 1944 aucun règlement officiel français ne traduisait de façon rationnelle et satisfaisante les effets réels du vent sur les constructions. Seul le Ministère de l'Air dans ses Normes relatives aux charges et surcharges pour projets d'exécution des Travaux Immobiliers de l'Aéronautique prescrivait une loi en "cos 2 α" donnant soit des pressions, soit des succions suivant l'exposition des surfaces. A la demande du Ministère de la Reconstruction une Commission, comprenant des spécialistes de l'Aérodynamique, de la Météorologie et de la Construction, fut créée pour élaborer des Règles tenant compte des données scientifiques et statistiques connues à l'époque.

Or, il y a vingt ans, ces données étaient limitées pour les bâtiments à des essais aérodynamiques surtout étrangers, ne couvrant qu'un domaine restreint et, pour les vitesses du vent en France, à l'expérience des techniciens de la Météorologie Nationale dont, après la guerre, les archives n'étaient pratiquement pas exploitables. Néanmoins, et non sans scrupules, la Commission entreprit une rédaction d'après ces renseignements incomplets et parfois contradictoires, pour répondre au souci de mettre rapidement à la disposition des utilisateurs un document (Bib. 1) permettant de faire face à la tâche considérable de la reconstruction dans des conditions alliant la sécurité et l'économie.

Conscients des lacunes et de certaines incertitudes de ce texte, les rédacteurs avaient envisagé la nécessité d'entreprendre des recherches et des études systématiques pour compléter et préciser les bases météorologiques et aérodynamiques d'une révision des Règles après quelques années d'application. Une enquête auprès des utilisateurs devait en outre faire ressortir en 1955 les améliorations de forme et de présentation souhaitables à l'occasion de cette révision.

1. PROGRAMME D'ETUDES & DE RECHERCHES

Le programme des recherches (Bib. 2 et 3) indispensables pour rendre le règlement cohérent, sûr et facilement applicable prévoyait des investigations dans les domaines aérologiques et aérodynamiques afin d'obtenir une connaissance plus précise des phénomènes climatiques et plus approfondie des faits physiques.

A. *Bases Météorologiques*

Un des objectifs primordiaux d'un règlement relatif à l'action du vent sur les constructions est la détermination de la pression dynamique à adopter pour une localité, un site et une construction donnés.

Dans un pays d'étendue relativement petite et sans relief important, comme les Pays-Bas ou la Belgique, il paraît possible de définir une ou deux vitesses de vent applicables à l'ensemble du territoire. Mais pour la France, avec ses plaines, ses côtes, ses vallées orientées comme le sillon rhodanien, ses massifs montagneux d'altitude variée, le réseau anémométrique ne paraissait pas assez dense pour permettre d'établir une carte des pressions dynamiques analogue par exemple à celle des prescriptions A.S.A. des Etats-Unis. Nous espérions voir unifier et compléter le cas échéant ce réseau français d'anémomètres au point de vue type d'appareil et mode d'enregistrement. Et, pour rechercher et obtenir des résultats absolument comparables entre eux, parvenir à affecter à chacune des stations météorologiques un coefficient de site approprié à la suite d'une étude du lieu d'érection y compris son environnement.

Nous souhaitions aussi procéder à une étude plus poussée de la structure des vents et de la turbulence, pour essayer d'en codifier les résultats et les transformer en règles d'application facile.

Enfin, les constructions se développant dans les pays francophones d'Afrique et les demandes relatives à la force et la nature des vents dans ces régions se multipliant, nous aurions voulu obtenir des données aussi précises que possible sur ces questions. Les effets particuliers des tornades auraient dû être examinés avec méthode pour y parer soit par une modification éventuelle des Règles afin de les adapter en vue d'une extension d'application à ces pays, soit par l'étude des dispositions d'ossatures ou de structures résistant mieux à ces vents spéciaux ou limitant les destructions à des éléments secondaires de la construction.

B. *Bases Aérodynamiques*

L'interprétation des effets du vent sur une construction réelle est délicate et les bases aérodynamiques correspondantes doivent avoir pour fondement des études et essais sur modèle réduit ainsi que des observations et mesures sur des constructions en vraie grandeur.

(a) - Etudes sur modèle réduit

Le programme des essais prévus était le suivant:

- vérification des diagrammes de pressions relatifs aux toitures isolées à un ou deux versants plans ou courbes.

- transition des toitures isolées aux constructions fermées par addition de parois verticales de hauteur croissante sous les toitures.

- influence de l'élancement, de l'allongement et de la profondeur des bâtiments.

- influence de la distance au sol.

- diagramme des pressions et détermination de la force d'entrainement sur des toitures multiples.

- définition de la rugosité des constructions cylindriques en fonction du diamètre.

- effets des pulsations et des changements brusques d'orientation.

Pour ce dernier point on avait envisagé de simuler en soufflerie les variations de vitesse en utilisant une hélice dont le pas pouvait varier rapidement et les variations d'orientation par des changements d'angle de la maquette par rapport aux axes vertical et horizontal, le tout selon telle cadence convenant à l'opérateur.

(b) - Etudes en vraie grandeur

Pour ces études sur des constructions en vraie grandeur, nous pensions recueillir des renseignements de deux manieres:

La première consistait, après chaque ouragan ayant produit d'importants dégâts, à photographier tout de suite ceux-ci et par une enquête, à essayer de déterminer les conditions dans lesquelles ils s'étaient produits.

La deuxième, à équiper une construction réelle d'appareils de mesure des déformations, des éléments de la structure et de prises de mesure des pressions (ou dépressions) locales dues à l'effet du vent dont la vitesse serait mesurée en synchronisme avec l'enregistrement des déformations et des pressions, puis de comparer le cas échéant, la distribution des pressions à celle obtenue au tunnel aérodynamique sur une maquette de la construction.

2. RESULTATS

A. Au point de vue aérologique et pour la France, c'est seulement à une date récente que nous avons commencé à travailler sur des données plus complètes et plus sûres. Nous recueillons les anémogrammes de ces dix dernières années correspondant aux plus grandes vitesses obtenues dans chaque station pour les dépouiller systématiquement d'après la méthode suivante:

Sur chaque anémogramme nous traçons:

- la courbe moyenne des vitesses, pondérée entre les valeurs maximales et minimales des pointes enregistrées,

- la courbe enveloppe des pointes de rafale (maximum absolu) en éliminant toutefois les pointes correspondant à une trop courte durée.

La première courbe sert à évaluer la vitesse correspondant à une pression dynamique dite normale; la deuxième est utilisée pour définir une pression dynamique exceptionnelle.

Cette étude statistique des enregistrements, n'est pas suffisamment avancée pour que nous puissions envisager le tracé d'un réseau de lignes d'égale vitesse maximale normale ou maximale exceptionnelle (ou des lignes d'égale pression dynamique correspondant à ces vitesses) choisies selon un échelonnement convenable.

Pour les pays francophones d'Afrique nous n'avons pas pu aborder la question car nous ne possédons pas davantage de renseignements qu'en 1946. Il en est de même en ce qui concerne la turbulence et la structure des vents.

B. Au point de vue aérodynamique le programme fixé a été presque complètement rempli (Bib. 4 - 5 - 6 - 7 - 8 - 9), sauf pour la transition des toitures isolées aux constructions fermées et les effets des pulsations et changements brusques d'orientation en ce qui concerne les essais sur maquette au tunnel.

Des essais complémentaires se poursuivent depuis le début de l'année 1963 pour obtenir une meilleure connaissance de la distribution des pressions et des forces d'entrainement sur des toitures multiples à deux versants égaux ou en sheds.

Pour les études en vraie grandeur, les tentatives faites pour tirer des enseignements valables des dégâts constatés après un ouragan ont donné peu de résultats positifs. Nous en avons retenu seulement le danger présenté par une fixation insuffisante des toitures légères et des éléments secondaires déduit de la fréquence des arrachements constatés le long des arêtes (rives - crêtes ou faitières de murs ou de toiture) et des corniches. Certains accidents survenus en cours de construction par exemple à des murs de pignon montés sans être contreventés ont attiré également l'attention de la Commission.

Depuis deux ans à l'Aéroport d'Orly un hangar est équipé pour enregistrer les forces, contraintes et pressions sur deux fermes principales en charpente métallique. Nous ne disposons pas encore de résultats suffisants pour en tirer des conclusions (Bib. 10).

3. LES REGLES NV 63

La rédaction des Règles NV 63 tient compte de tous les résultats obtenus au cours des essais. Elle marque en outre un important progrès par rapport aux Règles NV 46 qui nécessitaient parfois de se reporter à différents articles pour l'étude d'un type de construction. Le plan a été complètement repris, les différentes questions sont nettement séparées et pour chaque type de construction on retrouve toujours les mêmes paragraphes. La consultation et l'emploi des règles en sont facilités. Dans un esprit plus général la Commission a cherché à préparer, dans une certaine mesure, une future adaptation des Règles NV 63 aux notions de probabilité et d'état-limite pour pouvoir les mettre en harmonie avec d'autres règlements de construction tenant comptede ces notions. En effet, il semble que si le projeteur connaissait complètement la fréquence des vents d'une certaine force dans un lieu donné, leurs effets, compte tenu de l'environnement, sur une construction d'un type donné et le comportement de celle-ci à toutes les étapes qui conduisent à l'état limite de ruine, il deviendrait possible, en théorie, de calculer la probabilité, soit pour qu'elle ne subisse aucun dommage sous des surcharges survenant une ou plusieurs fois dans une année, soit pour qu'un élément essentiel de sa structure ne risque pas d'être mis hors d'état de remplir la fonction pour laquelle il a été conçu pendant toute la durée de cette construction.

Pour ne pas alourdir le présent exposé nous supposons que les Règles NV 46 sont connues des auditeurs et nous signalerons seulement les compléments ou modifications qui différencient les Règles NV 63 des précédentes.

L'Article 1 fournit tous les renseignements utiles à la compréhension et à une utilisation correcte des articles suivants. Il comporte l'ensemble des définitions et principes généraux indispensables ainsi que des indications sur la valeur des pressions dynamiques normales et exceptionnelles pour chacune des régions envisagées ainsi que la loi de variation de ces pressions avec la hauteur (*fig.1* à titre de rappel des Règles NV 46). Il est rédigé suivant l'ordre dans lequel se présentent en pratique les déterminations successives à faire pour calculer les forces dues au vent appliquées à une construction ou à un élément de cette construction. A ce titre il constitue une sorte de guide.

Conformément aux observations faites in situ sur les lignes hautes tensions les Règles NV 63 introduisent un coefficient de réduction des surcharges d'ensemble pour les grandes dimensions (*fig.2*). Les valeurs de ce coefficient découlent d'une adaptation des prescriptions belges en vigueur depuis 1960 (Bib. 13). Par contre les actions localisées, en particulier dans les zones critiques, sont l'objet de majorations.

Malgré leur importance, notamment pour les cheminées, phares, tours de télévision, pylônes-clochers, etc, ainsi que pour les ponts et toitures suspendues, les phénomènes d'accélération, de répétition et d'oscillation n'ont pas paru pouvoir être traduits par les Rédacteurs en règles précises, faciles à appliquer et couvrant tous les cas. Les Règles se bornent à attirer l'attention des ingénieurs et des constructeurs sur ces phénomènes.

L'Article 2 est relatif aux constructions à base rectangulaire en contact ou non avec le sol, couvertes par une toiture en terrasse, à un, deux ou plusieurs versants plans, ou en voûte cylindrique, cette toiture pouvant être unique ou multiple.

Un coefficient γ caractérise la construction en fonction du rapport de ses trois dimensions (*fig.3*). Il permet de définir assez correctement l'action (Bib. 6) des trois paramètres aérodynamiques principaux - élancement, allongement, profondeur - sur la valeur des coefficients. Il intervient de façon primordiale pour la détermination des actions extérieures et intérieures. Des essais sytématiques (Bib. 7) ont permis d'établir des courbes moyennes des coefficients de pression applicables aux toitures à 2 versants plans *(fig.4)* et aux toitures en voûte *(fig.5)* en fonction du coefficient γ. D'autres essais (Bib. 7) sur l'influence de l'éloignement bâtiment-sol ont permis de déterminer des valeurs limites pratiques pour les coefficients résultants de bâtiments rectangulaires en plan ainsi que pour les coefficients de pression applicables à la paroi inférieure parallèle au sol.

Bien que l'utilisation de cet article ne présente aucune difficulté, la Commission, pour plus de commodité et de rapidité a rédigé des prescriptions simplifiées et groupées pour tous les cas courants de bâtiments à base rectangulaire qui en fait représentent 90% de la totalité des constructions. Il convient de noter toutefois que cette simplification, fondée sur la notion d'enveloppe, est parfois acquise au prix d'une légère majoration des forces appliquées.

Les constructions prismatiques a base polygonale régulière ou circulaire, en contact ou non avec le sol, font l'objet de l'article 3. Elles sont classées en cinq catégories en fonction des essais aérodynamiques (Bib. 8 et 9) qui ont été effectués *(fig.6)*. Des échelles fonctionnelles permettent de lire le coefficient γ en fonction du rapport des dimensions et de tenir compte ainsi de l'élancement de la construction. Les coefficients de pression locaux à prendre en considération dans une section diamétrale sont donnés par des diagrammes en fonction de l'angle d'inclinaison de la surface en chaque point sur la direction du vent.

L'Article 4 concerne les panneaux pleins et toitures isolées.

- Panneaux: Le vent pouvant, sauf exception, attaquer un panneau plein sous tous les angles on a combiné dans une seule échelle fonctionnelle l'influence du rapport des dimensions et l'effet d'incidence oblique. Sous une telle incidence l'action totale peut en effet être supérieure à l'action donnée par une incidence normale pour certains élancements (valeur maximale pour la plaque carrée attaquée à 35°).

- Toitures: Les Règles NV 63 font intervenir le rapport des dimensions des toitures à 1 et 2 versants (Bib. 4), rapport non utilisé dans les Règles NV 46. Dans un but de simplification, seules les actions résultant des pressions ou dépressions sur les deux faces ont été données dans les règles *(fig.7)*. Il est en effet presque toujours inutile de connaître la valeur des pressions sur chacune des faces.

Le cas des toitures successives accolées, isolées dans l'espace sans aucun mur périphérique a été également examiné dans les Règles NV 63.

L'Article 5 traite des constructions ajourées et des constructions en treillis.

En général la rédaction est restée proche de celle des Règles NV 46. Toutefois pour des ensembles prismatiques rectangulaires ou triangulaires, tels que des tours ou pylônes en treillis, les coefficients globaux donnés dans les Règles NV 46 étaient trop élevés car ils couvraient les cas limites. La Commission de rédaction a tenu compte d'une récente étude belge et des essais exécutés par Joukoff et Tonglet sous la direction de Vandeperre. Elle a introduit une variation des coefficients en fonction du pourcentage de pleins. Les Règles serrent ainsi les résultats aérodynamiques (Bib. 11 et 12) de plus près.

A la demande des constructeurs métalliques une adaptation des tableaux belges (Bib. 13) et suisses des coefficients aérodynamiques relatifs aux profilés isolés a été donnée en Annexe.

Comme dans les Règles NV 46, l'article 6 des Règles NV 63 concerne les constructions diverses qui n'entrent pas strictement dans le cadre de celles définies dans les articles 2 - 3 - 4 - 5 **précédents**.

Des essais ont été réalisés pour déterminer les coefficients applicables à des coupoles reposant sur le sol ou sur une terrasse, et à des coupoles couronnant un cylindre. Il a été constaté que les forces ascendantes sont prépondérantes et très supérieures en valeur absolue aux forces horizontales.

Une formule permettant de calculer la force transmise par un drapeau en tissu à sa hampe a été tirée d'essais effectués par le Service de Recherches de l'Aéronautique.

Les Règles NV 63 comportent en outre des Annexes dont le rôle est:

(a) de compléter le règlement sur certains points tels que la détermination:

- de la hauteur fictive d'une construction située sur un terrain présentant des dénivellations importantes afin de pouvoir calculer la pression dynamique à utiliser,

- des actions intérieures dans les constructions comportant des parois partiellement ouvertes,

- des actions résultantes sur les parois d'une construction comportant plusieurs parois ouvertes,

- des actions normales du vent sur les éléments plans des constructions à treillis par sommation des forces appliquées à chacune des barres constitutives (profilés ou tubes) du treillis,

- de l'influence du rapport des dimensions des éléments plans uniques des constructions ajourées ou en treillis sur le coefficient global de trainée,

(b) de préciser l'application des règles au moyen d'un ensemble d'exemples de détermination des coeficinets de pression relatifs aux actions extérieures, intérieures et résultantes pour quelques types de construction.

4. CONCLUSIONS

Les Règles NV 63 réalisent un progrès important par rapport aux Règles NV 46, notamment en ce qui concerne les bases aérodynamiques. Par ailleurs leur nouvelle présentation en facilite l'emploi par les bureaux d'études.

Ce travail reste imparfait par suite de bases météorologiques encore insuffisantes et des effets peut-être mal appréciés des sollicitations du vent sur des constructions réelles en particulier pour les actions dynamiques et pulsations.

Aussi nous sommes heureux de féliciter le National Physical Laboratory de l'initiative qu'il a prise en réunissant à cette conférence des chercheurs scientifiques qui essaient d'établir les bases fondamentales de l'action du vent et des ingénieurs aux prises avec les problèmes pratiques. Nous espérons que cette liaison entre les constructeurs, les laboratoires et les offices météorologiques se poursuivra sur le plan international et conduira à des résultats fructueux pour tous. Le Comité W23 du Conseil International du Bâtiment déjà jumelé avec le Comité Europeen du Béton et l'ISO pourrait sans doute recevoir des renseignements intéressants d'un Centre de documentation et de coordination relatifs à l'action du vent sur les constructions qui continuerait l'oeuvre d'information et d'échange de vues entreprise par cette conférence.

A titre de simple suggestion nous pourrions envisager comme programme:

Au point de vue aérodynamique:

- de normaliser les méthodes d'essais au tunnel afin d'obtenir pour une même forme de construction des valeurs comparables dans toutes les souffleries.

- de se communiquer les résultats d'essais obtenus dans les laboratoires des différents pays et éventuellement de commencer par réunir et analyser ceux déjà réalisés dans le passé.

Au point de vue météorologique:

- d'harmoniser dans tous les pays les méthodes de mesure de la vitesse du vent.

- de définir un mode d'utilisation des anémogrammes pour obtenir la ou les vitesses de base applicables au calcul des constructions.

- d'établir une échelle internationale des vitesses et de s'entendre sur la vitesse à adopter dans les régions frontières entre des pays limitrophes.

Au point de vue construction:

- d'observer les effets du vent sur des constructions réelles pour en déduire des règles conformes à leur comportement.

LISTE des FIGURES

TITRES & LEGENDES

Les figures, telles qu'elles sont reproduites dans la présente communication, correspondent à la rédaction actuellement en cours d'examen par différents organismes officiels en vue de l'homologation des Règles NV 63. Elles sont donc susceptibles d'être modifiées comme présentation ou d'être adaptées à des changements de texte.

1. Région I - Site normal - Vent normal et exceptionnel. Loi de variation de la pression dynamique avec la hauteur

Pour H compris entre 0 et 500 m, le rapport entre la pression dynamique q_H à la hauteur H et la pression dynamique q_{10} à la hauteur standard de 10 m est donné par la formule:

$$\frac{q_H}{q_{10}} = 2,5 \frac{H + 18}{H + 60}$$

2. Influence des dimensions sur les actions du vent

Ce diagramme n'est pas applicable à des parties de construction situées à une cote supérieure à 50 m par rapport au terrain environnant. Les coefficients de réduction font actuellement l'object d'un nouvel examen dans le sens de la prudence.

3. Constructions rectangulaires - Coefficient γ

Entre autres ce coefficient γ permet de choisir la loi de variation des coefficients de pression applicables à une toiture un vent normal au faitage *(fig.4)* ou à la génératrice de clef *(fig.5)*. Pour une paroi verticale appartenant à une construction fermée le coefficient de pression C_e, lorsque le vent agit perpendiculairement à cette paroi a pour valeur:

Face au vent $C_e = + 0,8$

Face sous le vent $C_e = -(1,3\ \gamma - 0,8)$

4. Constructions rectangulaires - Toitures à un ou plusieurs versants plans

5. Constructions rectangulaires - Toitures en voûte

6. Constructions prismatiques à base polygonale régulière ou circulaire

   (a) - Tableau du coefficient global de trainée c_t

   (b) - Echelle fonctionnelle du coefficient γ

λ est le rapport de la hauteur (ou de la longueur) de la construction à la largeur de son maitre-couple.

7. Toiture isolée à deux versants plans et symétriques

λ est défini conventionnellement dans les Règles: il dépend essentiellement du rapport de la dimension d'un versant suivant la ligne de plus grande pente à la dimension horizontale parallèle à l'arête du dièdre.

BIBLIOGRAPHIE

1. Règles définissant les effets de la neige et du vent sur les constructions applicables aux travaux dépendant du Ministère de la Reconstruction et de l'Urbanisme et aux Travaux privés dites "Règles NV 46", janvier 1947.

    Editeur: Société de Diffusion des Techniques du Bâtiment et des Travaux Publics - 9, rue La Pérouse - Paris 16è.

2. ESQUILLAN, N. - Les nouvelles règles françaises relatives à l'action de la neige et du vent sur les constructions.

    Circulaire Série I N° 38 du 8 novembre 1947
    Institut Technique du Bâtiment & des Travaux Publics
    9, rue La Pérouse - Paris 16è.

    Cette Circulaire est suivie d'une bibliographie correspondant aux principaux documents utilisés pour la rédaction des Règles NV 46.

3. ESQUILLAN, N. - Application des Règles édictées en 1946 par le Ministère de la Reconstruction et de l'Urbanisme

    Actes du Colloque International de Mécanique -
    Poitiers 1950 Tome III - Etudes sur la Mécanique des Fluides

    Publications Scientifiques et Techniques du Ministère de
    l'Air N° 251 - 1951

4. Etudes aérodynamiques concernant l'action du vent sur les toitures isolées

    Efforts globaux - Efforts locaux - Influence de l'allongement
    Cahiers du Centre Scientifique et Technique du Bâtiment N° 21 -
    Cahier 195 - 1954

5. PRIS, R. - Action du vent sur les bâtiments et constructions, préparation et montage des maquettes en soufflerie

    Annales ITBTP Février 1962 N° 170
    Série: Essais et mesures (54)

6. PRIS, R. - Bâtiments rectangulaires en plan avec toitures terrasses. Détermination des forces et des pressions dues à l'action d'un vent normal aux faces des bâtiments. Etablissement du diagramme des coefficients résultants

   Annales ITBTP Janvier 1963 N° 181
   Série: Essais et mesures (62)

7. PRIS, R. - Détermination des pressions dues à l'action du vent sur les toitures des bâtiments rectangulaires en plan en contact avec le sol. Toitures à deux versants. Toitures cylindriques. Détermination des pressions sur des toitures sphériques. Bâtiments éloignés du sol. Influence de l'éloignement sur les caractéristiques aérodynamiques de bâtiments rectangulaires en plan

   Annales ITBTP, 1963
   Série: Essais et mesures (en cours de publication)

8. PRIS, R. - Réservoirs et cheminées. Recherches préliminaires Influence de la rugosité et de la nervuration sur la résistance aérodynamique des cylindres

   Annales ITBTP Novembre 1960 N° 155
   Série: Essais et mesures

9. PRIS, R. - Réservoirs et cheminées. Résistance de cylindres à base polygonale à arêtes vives ou munis de nervures. Résistance de cylindres à base circulaire et à surface rugueuse

   Annales ITBTP Juillet-Août 1961 N° 163-164
   Série: Essais et mesures (51)

Rapports N° 8551 - 1 & 2

   Etudes expérimentales de l'action du vent sur les éléments d'échafaudages tubulaires

   Institut de Recherches et d'essais du Centre Ouest Poitiers 1951

PRIS, R. - Etude aérodynamique d'une tour de réfrigération hyperbolique

   Annales ITBTP Février 1959 N° 134
   Série: Essais et mesures (42)

FISCHER, A. - Construction des grandes tours de réfrigérants

Problèmes théoriques et essais
Résultats des essais sur maquette en soufflerie

Construction   Tome XIV - N° 10   Octobre 1959
*   Dunod éditeur*

10. Compte rendu de recherches effectuées en 1961 par les organismes de l'Union Interfédérale du Bâtiment et des Travaux Publics.

    Mesures des contraintes subies par les bâtiments de grande envergure sous l'action du vent

    Annales ITBTP   Octobre 1962   N° 178
    Série: Questions Générales (58)

11. JOUKOFF. - Action du vent sur les pylônes de section triangulaire

    L'Ossature Métallique   Novembre 1950
    Centre Belgo-Luxembourgeois d'information de l'acier

12. VANDEPERRE & NAAR, Jacques. - Action du vent sur les pylônes en treillis

    d'après les Essais de M. Tonglet en 1952

    Annales des Travaux Publics de Belgique   N° 5-1956

13. Normes Belges relatives à l'action du vent sur les constructions   1960

    NBN 160-01   Instructions générales pour le calcul de l'action du vent sur les constructions

    NBN 160-02   Pièces longues droites pleines ou en treillis à arêtes vives

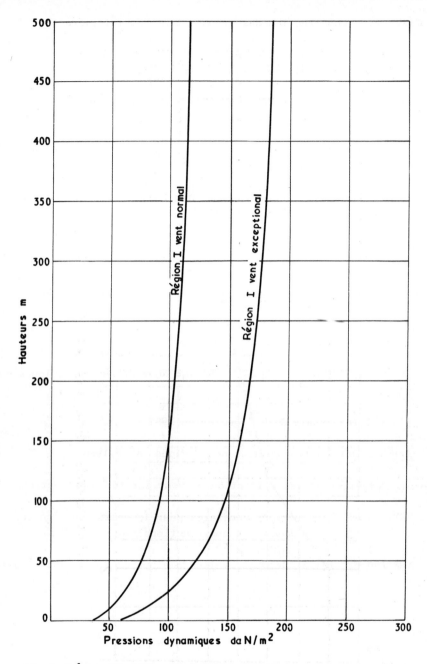

Fig.1. Région I - Site normal - Vent normal et exceptionnel
Loi de variation de la pression dynamique avec la hauteur

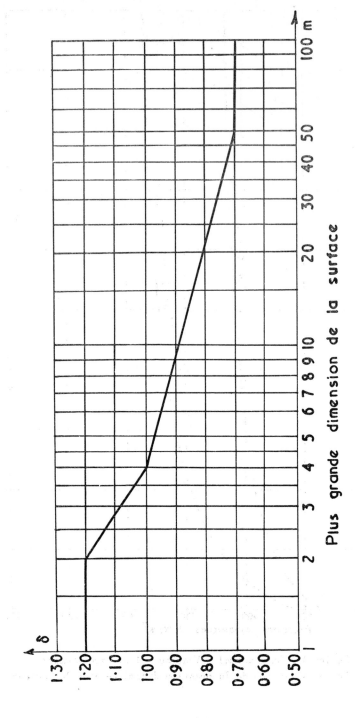

Fig.2. Influence des dimensions sur les actions du vent

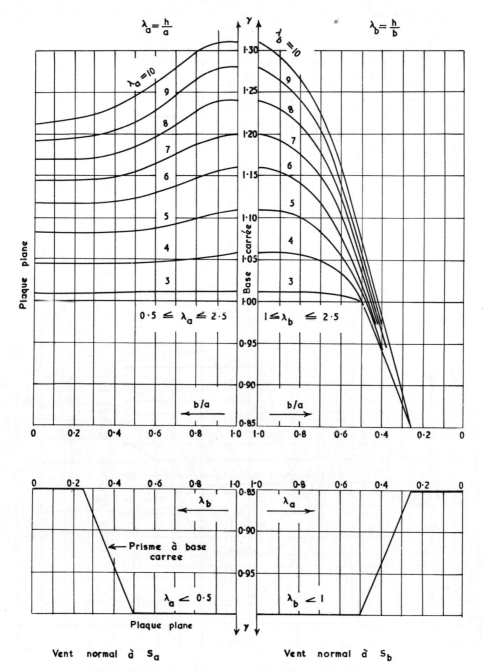

Fig.3. Constructions rectangulaires - Coefficient γ

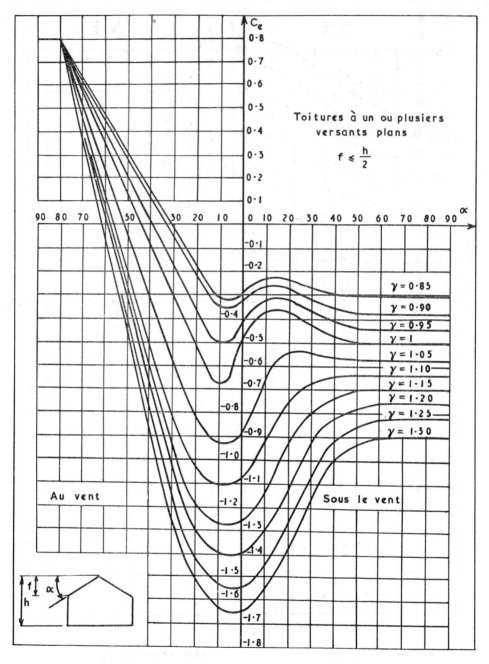

Fig.4. Constructions rectangulaires - Toitures à un ou plusieurs versants plans

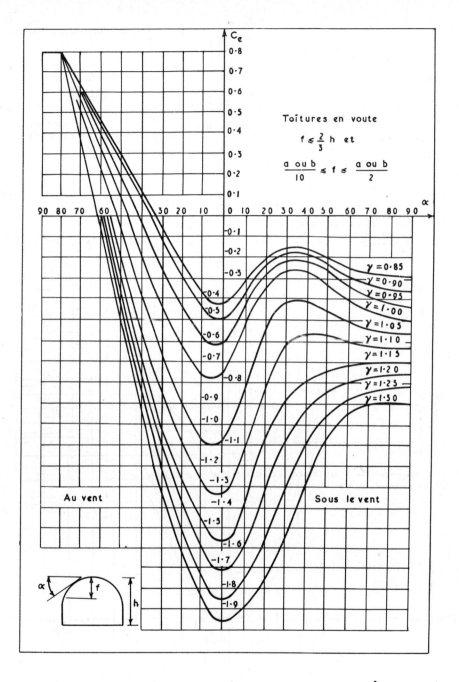

Fig.5. Constructions rectangulaires - Toitures en voûte

Catégorie	Nature de la construction		Coefficients	
			c_{t0}	$c_{t\infty}$
I	Prismes de 3 ou 4 côtés		1·30	2·00
II	Prismes de plus de 4 côtés et de moins de 12 côtés avec ou sans nervures arrondies	5 côtés	1·05	1·62
		6 côtés	1·00	1·54
		8 côtés	0·95	1·46
		10 côtés	0·85	1·31
III	Prismes de 12 côtés et plus, avec ou sans nervures vives ou arrondies		0·80	1·23
IV	Cylindres à base circulaire avec nervures minces ou épaisses à arêtes vives (saillie comprise entre 0·01 et 0·10)		0·75	1·16
V	Cylindres rugueux à base circulaire sans nervure.			
	$d\sqrt{q} \leq 0.5$ ou $d \leq 0.04$		0·75	1·00
	$0.5 < d\sqrt{q} < 1.5$		$0.85 - 0.2\, d\sqrt{q}$	$1.135 - 0.27\, d\sqrt{q}$
	$d\sqrt{q} \geq 1.5$ ou $d \geq 0.28$ [1]		0·55	0·73

(1) Dans ces inégalités d est exprimé en mètres, q en decaNewtons par mètre carré et V en mètres par seconde.

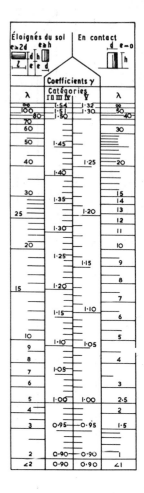

(a) Tableau du coefficient global de trainée c_t

(b) Echelle fonctionelle du coefficient γ

Fig.6. Constructions prismatiques à base polygonale régulière ou circulaire

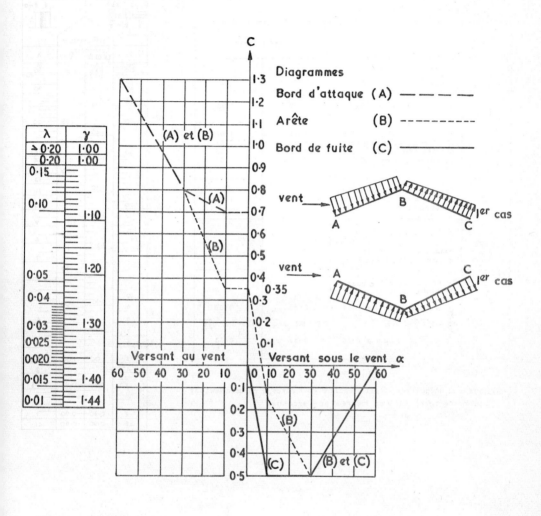

Fig.7. Toiture isolée à deux versants plans et symétriques

PAPER 14

PROPOSED CODE OF PRACTICE FOR WIND LOADS FOR DENMARK

by

MARTIN JENSEN and NIELS FRANCK
(Denmark)

PROPOSED CODE OF PRACTICE FOR WIND LOADS FOR DENMARK

by

MARTIN JENSEN AND NIELS FRANCK

THE new Danish Code for Wind Loads was published for comment in June 1962. From the letter that accompanied the publication can be mentioned:

Motive

An accident in Pødovre in 1956 when some large roofs were blown down during a storm initiated investigations by the Wind Laboratory of The Technical University of Denmark on suction on roofs with small slopes.

These investigations showed that the suction was far greater than was required in the existing code.

Preliminary supplement to the code

The working committee immediately produced a supplement to the existing code which proposed a considerable increase of the suction on roofs with small slopes.

The Institution of Danish Civil Engineers incorporated this supplement in the code in 1956.

Experiments

It was clear to the Committee that the new code for wind loads must be based on new laboratory experiments.

Since Irminger and Nøkkentved carried out the model experiments that are the basis for the existing code the Wind Laboratory has developed a new model law and corresponding technique that gives results in much better accordance with conditions in nature. A preliminary report on this can be found in Ingeniøren, International Edition, vol. 2, No. 4, 1958; Martin Jensen: The Model-Law for Phenomena in Natural Wind.

The Committee therefore resolved in 1956 that investigations both in wind tunnel and nature should be carried out.

These investigations were performed in the Wind Laboratory by Niels Franck and Martin Jensen.

The investigations were completed at the end of 1960.

In addition to the Danish experiments a selection of foreign investigations were used for the Proposal.

Maximum wind velocities

In 1957 three stations for wind measurements were established. In the stations was measured, inter alia, the velocity pressure of the wind.

A statistical computation of this material until Summer 1961 has been published. **Martin** Jensen: Maksimale Vindhastigheder i Danmark, Ingeniøren nr. 23, December 1961. These results are the basis for the section on wind velocities in the Proposal.

However, measurements are being continued and a final result can be expected in about 3 years.

The proposal

The Proposal that follows is considerably more detailed than the existing code. This has been possible because the new model experiments have furnished substantially better knowledge about wind loads than formerly existed.

The detailed specifications enable the design engineer to proportion the structure in close accordance with the loads to which it will be exposed. It should be stressed that simplified specifications imply a systematic over-dimensioning of the structure.

To introduce the possibility for economic construction the Committee has considered it sound to give rather detailed specifications. In cases where it is unreasonable to carry out detailed calculations the design engineer himself can make out, on the basis of the detailed specifications, a simplified set.

Dynamic effects of the wind

Under preparation is a 4th part of the code dealing with the dynamic effects of the wind on structures.

INTRODUCTION

0.1 The wind load is the product of a shape factor, an area and the velocity pressure of the wind.

0.2 Normally the wind load can be considered to act at right angles to the surface in question either as a pressure i.e. directed against the surface or as a suction i.e. directed away from the surface.

In special cases the tangential wind load may be of some importance.

0.3 The wind load is a live load. Normally it will be sufficient to consider $\frac{1}{3}$ of the wind load as live load.

0.4 The load factor f_v for the wind load can be taken as:

$f_v = 1.1$ in cases where the shape factors are determined in a carefully performed model test which completely represents the actual conditions.

$f_v = 1.2$ in cases where the shape factors noted in this code are used.

$f_v \geq 1.5$ in cases where the shape factors are determined with reasonable judgement.

1. WIND VELOCITY

The wind velocity increases with the height over the terrain following a logarithmic function.

The maximum wind velocities given in 1. appear with a probability of the order of 1/500 per year.

1.1 Normal velocity profiles.

1.1.1 For Denmark excluding Greenland the maximum wind velocities in open terrain can be put at:

$v = 27 (1 + \frac{1}{2}\log_{10} z)$, v is the wind velocity in m/s and z is the height over the terrain in m.

For places in Jutland lying less than 10 km from the North sea the sea winds must be calculated at 1.1 times greater velocities.

1.1.2 For Greenland the maximum wind velocities in open terrain can be put at:

$v = 36 (1 + \frac{1}{2}\log_{10} z)$, v is the wind velocity in m/s and z is the height over the terrain in m.

For places in Greenland lying less than 10 km from a coast exposed to winds from the Atlantic Ocean, the Davis Strait or the Baffin Bay the sea winds must be calculated at 1.1 times greater velocities.

At places in Greenland where foehn winds occur for the wind directions considered one must, according to circumstances, calculate for greater wind velocities than those considered above.

1.2   Exposed and protected areas.
1.2.1   For places where the conditions of the terrain are judged to give extraordinarily strong winds, one must, for the directions in question take greater wind velocities than those given in 1.1.
1.2.2   For calculation of structures that will be built on places sheltered from the wind, smaller wind velocities than those given in 1.1 can be used. It is assumed that the shelter will exist for as long as the structure.

If the wind considered passes more than 0.5 km of built-up area or forest, structures less than 30 m high can be calculated for wind velocities 0.8 times those in 1.1.

1.3   Reduced wind velocities.
1.3.1   In cases of short duration such as temporary constructions and situations during erection smaller wind velocities than those in 1.1 and 1.2 can be used.
1.3.2   In cases where the wind load is combined with other loads which according to the nature of things will not coincide with high wind velocities, smaller wind velocities than those in 1.1 and 1.2 can be used.

This applies for example to calculations for operating conditions of cranes which can be assumed inoperative during high wind velocities.

1.3.3   For combinations of wind loading and another loading which occur extremely rarely smaller wind velocities than those in 1.1 and 1.2 can be used.

This applies for example to a combination of wind loading and earthquake.

2.   VELOCITY PRESSURE

The force of the wind is proportional to the kinetic energy $q = \frac{1}{2}\rho v^2$, where ρ is the mass per unit volume of the air and v is the wind velocity. If one inserts $\rho = \frac{1}{8} \frac{kg}{m^3} \div \frac{m}{s^2}$ and v in m/s, q will have the dimension kg/m^2, q being called the velocity pressure, $q = \frac{1}{16} v^2$; v m/s, q kg/m^2.

3.   SHAPE FACTORS

Introduction and notation.

Normally the shape factors cannot be determined theoretically. Most of them are measured in scale model tests.

When shape factors are used that are determined by model experiments it must be assured that the model experiment represents the actual situation.

The shape factors given below can be used in the calculation of wind loads unless a more exact determination is preferred.

For structures that differ from the following examples it will usually be necessary to determine the shape factors experimentally.

The wind load is calculated from the following shape factors:

- c is the factor for the wind load on a surface or part of a surface, and is always positive since the direction of the wind force in every case is designated by "pressure" or "suction".
- $\bar{c}$ is the factor for the mean value of the wind loads over a surface or part of a surface, and is always positive since the direction of the wind force in every case is designated by "pressure" or "suction".
- c_t is the factor for the total wind load on a section i.e. for the combined action of the wind load on the two surfaces of the section, and is always positive, since the direction of the wind force in every case is designated.
- c_i is the factor for an interior pressure.
- c_τ is the factor for the tangential component of the wind load on a surface.
- C is the factor for the total wind force in the direction of the wind on a structure or part of a structure.
- C_e is the factor for the total wind force on a screen or a truss. The relevant area is the effective area of the structure.

3.1 Tangential wind load.

The tangential component of the wind load depends on the roughness of the surface. Usually it is insignificant. For example when the wind is blowing parallel to the surface the factor c_τ is: even concrete surface $c_\tau = 0.002$, corrugated sheets (asbestoscement) $c_\tau = 0.02$, surface with 25 cm projecting ribs $c_\tau = 0.04$, in the latter two examples for the wind at right angles to the corrugations or the ribs.

3.2 Bars, screens, trusses and bridges.

The wind load on a screen or a truss consisting of members with angular sections depends on the solidity $m = F_e/F$, where F_e is the effective area and F the area within the contour. The shape factor C_e applies to the effective area.

At some height above the ground level the wind can blow obliquely up or down. The influence of such wind movements in the vertical plane can be taken into account by varying the calculated wind force from an inclination of 0.2 below the horizontal to 0.2 above the horizontal.

3.2.1 **Bars**.
The wind load on prismatic bars with lengths in relation to cross sectional dimensions relatively large, can be calculated from the shape factor C given below. It is assumed that the wind blows at right angles to the length of the bar. The shape factor applies to the component of the wind force in the direction of the wind, and to the projected area of the bar on a plane normal to the direction of the wind.

The wind load on round bars depends on the Reynolds' number Re. For a smooth circular cylinder can be used:
C = 1.2 for Re < 5×10^5,   subcritical
C = 0.7 for 5×10^5 < Re, supercritical.
Re = $\frac{v \, d}{1.5} \, 10^5$, where v is the wind velocity in m/s, and d is the diameter in m.

A painted steel surface can be considered smooth.

For circular cylinders with rough surfaces C is greater, especially in the supercritical range.

For ropes C = 1.4 can be used.

The wind load for bars of angular section such as rolled steel sections, rectangles and the like corresponds somewhat to C = 2. If the cross section is not symmetrical to the wind there is in addition a cross force.

3.2.2 **Screens at ground level**.
If the wind blows at right angles to a screen which is standing on the ground then the shape factors given in figure 1 can be used. The relevant velocity pressure is calculated level with the top of the screen. The figure applies to screens with rectangular shapes, s is the ratio of the screen's length to its height.

3.2.3 **Single screen or truss above ground level**.
If the direction of the wind is normal to the screen or the truss and if there is free passage for the wind between the structure and the ground, the shape factors in figure 2 can be used. The figure applies to rectangular structures, t is the ratio between the longest and shortest sides.

If the direction of the wind is about 45° to a solid screen and if t is small then the wind force is greater than for an angle of attack of 90°. This increase is 10% for t = 3 growing to 50% for t = 1.

3.2.4 **Pair of trusses**.
The wind load on a structure consisting of two parallel, identical trusses can be calculated as follows:

The windward truss is calculated as in 3.2.3.

The wind load for the leeward truss depends on how exposed it is to the wind. This is expressed by the factor f.

Assuming that $0.2 < h/b < 1.0$ where h is the truss height and b is the distance between the two trusses, then f can be calculated from:

$m < 0.6$ $\qquad f = 1.15 - 1.67\, m\, \sqrt[4]{\dfrac{h}{b}}$

$0.6 < m$ $\qquad f = 1.15 - m\, \sqrt[4]{\dfrac{h}{b}}$

values of f greater than 1.0 are taken as 1.0.

The wind load V on the leeward truss is $V = C_e f\, q\, F_e$, where C_e is the shape factor for the windward truss.

3.2.5 Bridges.

The wind load on the superstructure of a bridge which has two identical main trusses consisting of angular elements can, when the wind is normal to the bridge, be calculated as given below.

The windward truss is calculated as given in 3.2.3.

The wind load on the deck assembly, traffic loads and leeward main-truss depends on how exposed they are to the wind. This is expressed by the factor f.

The deck assembly is given the shape factor $C = 2.0$. $f = 1 - m$ can be used for the part behind the windward truss's contour area and $f = 1$ for the remaining part.

The traffic load is given the shape factor $C = 2.0$. When the traffic load is lower than the windward truss $f = 1 - m$ is used. When the traffic load is higher than the windward truss $f = 1$ can be used for the part above the truss, and for the remaining part can be used $f = 1 - m$ when $m < 0.5$ or $f = 0.5$ when $0.5 < m$.

The height of the traffic load can be taken as:
on railways bridges 3.8 m,
on road bridges   2.5 m, and
on footbridges   1.7 m.

The wind load on the leeward main truss is calculated as given in 3.2.4. On the part of the leeward main truss which is behind the deck assembly $f = 0$. If the traffic load is placed close to the leeward main girder $f = 0$ can be used for the part which is behind the traffic load.

3.2.6 Lattice masts and lattice towers.

For the calculation of the wind load on lattice masts and lattice towers the velocity pressures at the respective heights are used.

For lattice masts and lattice towers with square cross-sections consisting of angular members the wind load, for the wind blowing normal to a side, can be calculated as given in 3.2.4. The wind load on the two trusses parallel to the wind is insignificant.

For the wind blowing at 45° to the sides the wind load is 1.2 times as great.

The wind load on lattice masts and lattice towers with triangular cross-sections consisting of cylindrical members can be calculated by assigning to the chords the shape factor C_{fl}, to the cross bars C_{gi} and to the fish-plates C_k as given below. These shape factors apply to the developed form of the structure.

$C_{fl} = 1.2$ \qquad\qquad $Re < 5 \times 10^5$
$C_{fl} = 0.7$ \qquad $5 \times 10^5 < Re$
$C_{gi} = 0.6$ \qquad\qquad $Re < 5 \times 10^5$
$C_{gi} = 0.35$ \qquad $5 \times 10^5 < Re$
$C_k = 0.5$
$Re = \dfrac{v\,d}{1.5} 10^5$, where v is the wind velocity in m/s and d is the diameter in m.

The wind load changes only a little when the wind veers.

3.3 Buildings.

The shape factors in section 3.3 apply to buildings with rectangular plans, vertical walls and with heights less than 3 times the greatest horizontal dimension. They can be used for buildings supported on columns if the free height under the building is less than a third of the height from the ground to the top of the building.

The building's height h is, unless otherwise noted, the distance from the ground to the top of the roof excluding small projections such as ventilation cowls, boxes for lift hoists or expansion tanks, chimneys and the like.

The length of the building l is measured parallel to the ridge and the width normal to the ridge. If the roof is horizontal l and b are respectively the longest and shortest dimensions.

The roof slope α is the tangent of the angle between the roof surface and the horizontal and is always positive.

All shape factors in section 3.3 apply to the velocity pressure level with the top of the roof.

3.3.1 Exterior walls.

Usually it will be sufficient to investigate the stability of a building for lengthwise and crosswise wind loads. The part of these loads resulting from the exterior walls can be calculated from the following shape factors:

pressure on the windward wall $\bar{c} = 0.7$
suction on the leeward wall $\bar{c} = 0.5 + 0.08\dfrac{h}{n} - 0.1\dfrac{p}{h} - 0.12\dfrac{p}{n}$,

but $0.1 < \bar{c}$; h is the height of the building, n is the dimension of the building normal to the wind and p is the dimension parallel to the wind.

A part of a windward wall with area f can receive the greater load on its exterior side the smaller f is in relation to the total wall area F.

The inward acting wind load on the outer surface of a part of a windward wall can be calculated from $\bar{c} = 1.0 - 0.3\frac{f}{F}$.

The wind load on the outer side of a wall that is parallel to the wind can be calculated from the following shape factors,
areas with distances less than h/2 from the windward edge of the wall, $\bar{c} = 1.2$ outwards,
remaining parts of the wall, $\bar{c} = 0.6$ outwards.

The wind load on the outer side of the leeward wall can be calculated from $\bar{c} = 0.5$ outwards.

The load on the inner side of the exterior wall is the same as the over or the underpressure c_i in the building.

If the building is predominantly permeable or completely open in the windward side there is an overpressure corresponding to $c_i = 0.7$. If the building is predominantly permeable or completely open in one of the two sides which are parallel to the wind $c_i = 0.8$ underpressure can be used. If the building is predominantly permeable or completely open in the leeward side $c_i = 0.5$ underpressure can be used. If the building is equally permeable in all 4 sides $c_i = 0.3$ underpressure can be used.

A side that has hatches, doors or windows that can be opened must usually be considered predominantly permeable in the most unfavourable way.

3.3.2 <u>Pitched roofs with or without hips</u>.

The shape factors apply to wind loading on the upper side of the roof.

When the wind blows normal to the ridge the shape factors given in figure 3 can be used.

When the wind blows skew to the building the shape factors given in figure 4 can be used. At the border n at the windward corner when the roof slope is below 0.4 there must, in addition to the suction $\bar{c} = 1.8$, be placed a movable section of dimensions 0.1b x 0.2b with an extra suction $\bar{c}_b = 1.8$.

For the wind blowing parallel to the ridge the shape factors given in figure 5 can be used.

3.3.3 <u>Monoslope roofs</u>.

The shape factors apply to the wind load on the upper side of the roof.

When the wind blows normal to the horizontal generators of the roof the shape factors given in *fig. 6* and *7* can be used.

When the wind blows skew to the building the shape factors given in *fig.8* and *9* can be used. At the border q at the low corner leading into the wind when the roof slope is below 0.4 there must, in addition to the suction $\bar{c}$ = 1.8, be placed a moveable section of dimensions 0.1b x 0.2b with an extra suction $\bar{c}_b$ = 1.8. At the border r at the high corner leading into the wind there must in addition to the suction $\bar{c}$ = 1.8, be placed a moveable section of dimensions 0.1b x 0.2b with an extra suction $\bar{c}_b$ = 1.8.

When the wind blows parallel to the horizontal generators of the roof the shape factors given in *fig.10* can be used.

3.3.4 <u>Cylindrical roofs, camber over 1/8.</u>

The shape factors apply to the wind load on the upper side of the roof.

When the wind blows normal to the generators of the roof the following shape factors can be used:
windward areas where $0.8 < \alpha$, pressure $\bar{c}$ = 0.7,
areas where $\alpha < 0.8$, suction $\bar{c}$ = 1.4,
leeward areas where $0.8 < \alpha$, suction $\bar{c}$ = 0.5.

When the wind blows skew to the building the wind load is not known. It is recommended to load a border of width 0.1b along the gable with a suction $\bar{c}$ = 1.8 and in addition to place on this border a movable section of dimensions 0.1b x 0.2b with an extra suction $\bar{c}_b$ = 1.8.

When the wind blows parallel to the generators the shape factors given for pitched roofs can be used.

3.3.5 <u>Cylindrical roofs, camber less than 1/8.</u>

It is recommended to use the shape factors given for pitched roofs, putting $\alpha = 2p/b$, where p is the rise of the roof and b is the width of the building normal to the roof's generators.

3.3.6 <u>Saw tooth roofs.</u>

The wind load on saw tooth roofs is only incompletely known.

When the wind blows normal to the ridges it is recommended to calculate the wind load on the upper surface of the roof by using the shape factors given in *fig.11*.

When the wind blows skew to the building the wind load is not known. There are probably large suctions on a border along the windward gable and on a border along the windward facade.

When the wind blows parallel to the ridges the shape factors given for monoslope roofs can be used.

3.3.7 **Underside of roofs.**

Underside of roof overhang:

Roof overhang at a windward wall, pressure on the underside $\bar{c} = 0.7$,
" " " " leeward " , suction " " " $\bar{c} = 0.5$,
" " " " wall parallel to the wind, suction on the underside $\bar{c} = 0.6$.

When the wind blows skew to the building, pressure on the underside of the roof overhang at the two windward walls $\bar{c} = 0.7$.

The wind load on the underside of a roof, possibly on the uppermost ceiling is the same as the over or under pressure c_i in the room underneath. If this room is equally permeable in all sides an under pressure $c_i = 0.3$ can be used. If the wind blows along or across the building and the windward side is predominantly permeable or completely open there is an overpressure corresponding to $c_i = 0.7$; if the leeward side or one of the sides parallel to the wind is predominantly permeable or completely open an underpressure $c_i = 0.6$ can be used. If the wind blows skew to the building and one of the two windward sides is predominantly permeable or completely open an overpressure $c_i = 0.6$ can be used; if one of the two leeward sides is predominantly permeable or completely open an underpressure $c_i = 0.5$ can be used.

A side that has hatches, doors or windows that can be opened must usually be considered predominantly permeable in the most unfavourable way.

3.4 **Free roofs.**

The shape factors given in 3.4 apply when there is free passage between the ground and a roof for a wind the direction of which is across the roof.

The length direction of the roof is the direction of the horizontal generators.

The shape factors given in 3.4 assume that the free height under the roof is over one half of the width of the roof measured horizontally.

The roof slope α is the tangent of the angle between the roof surface and the horizontal and is always positive.

The shape factors in 3.4 apply to the velocity pressure level with the uppermost generator.

The shape factor c_t applies to the total wind load i.e. combined action of the wind loads on the over and under sides of the roof. The load acts normal to the roof; the designations "upwards" or "downwards" only apply to the direction of the vertical component of the force.

3.4.1 **Pitched roofs.**

When the wind blows at right angles to the length of the roof the

shape factors given below can be used (see *fig.12*):
$\alpha < 0.3$
windward part of the roof upwards $\bar{c}_t = 1.0 - 2\alpha$
or downwards $\bar{c}_t = 1.0 + 2\alpha$
leeward part of the roof upwards $\bar{c}_t = 1.0$.
$0.3 < \alpha < 0.5$
windward part of the roof upwards $\bar{c}_t = 1.0 - 2\alpha$
or downwards $\bar{c}_t = 1.0 + 2\alpha$
leeward part of the roof upwards $\bar{c}_t = 2.5 - 5\alpha$.
$0.5 < \alpha$
windward part of the roof downwards $\bar{c}_t = 2.0$
leeward part of the roof $\bar{c}_t = 0$.

3.4.2 Trough roofs.

When the wind blows at right angles to the surface of the roof the shape factors given below can be used (see *fig.13*):
$\alpha < 0.3$
windward part of the roof downwards $\bar{c}_t = 1.0 - 2\alpha$
or upwards $\bar{c}_t = 1.0 + 2\alpha$
leeward part of the roof downwards $\bar{c}_t = 1.0$.
$0.3 < \alpha < 0.5$
windward part of the roof downwards $\bar{c}_t = 1.0 - 2\alpha$
or upwards $\bar{c}_t = 1.0 + 2\alpha$
leeward part of the roof downwards $\bar{c}_t = 2.5 - 5\alpha$.
$0.5 < \alpha$
windward part of the roof upwards $\bar{c}_t = 2.0$
leeward part of the roof $\bar{c}_t = 0$.

3.4.3 Monoslope roofs and horizontal roofs.

The wind load on free monoslope roofs can be considered as varying linearly from c_t^w at the windward edge to c_t^l at the leeward edge.

When the wind blows at right angles to the length of the roof the shape factors given below can be used when the high edge is leading into the wind (see *fig.14*):

$0 \leq \alpha < 0.2$ upwards or downwards $c_t^w = 2.0$ $c_t^l = \alpha$

$0.2 < \alpha < 2.0$ upwards $c_t^w = 2.0$ $c_t^l = \alpha$

$2.0 < \alpha$ upwards $c_t^w = 2.0$ $c_t^l = 2.0$.

When the wind blows at right angles to the length of the roof the shape factors given below can be used when the low edge is leading into the wind (see *fig.15*):

$0 \leq \alpha < 0.2$ downwards or upwards $c_t^w = 2.0$ $\quad c_t^l = \alpha$

$0.2 < \alpha < 2.0$ downwards $\quad c_t^w = 2.0 \quad c_t^l = \alpha$

$2.0 < \alpha \quad$ downwards $\quad c_t^w = 2.0 \quad c_t^l = 2.0$.

3.5 Towers and chimneys.

For the calculation of the wind load on towers and chimneys the velocity pressures at the respective heights must normally be used.

If the cross section is circular and the surface is even $C = 0.7$ can be used. The relevant area is the diameter times the height. This value can be used for masonry surfaces and for surfaces of riveted or bolted steel plate. If the surface is more uneven the shape factor will be greater than 0.7.

For towers and chimneys with rectangular cross sections it will normally be sufficient to investigate the stability for lengthwise and crosswise wind loads. The shape factors can be calculated from:
$C = 1.2 + 0.08\frac{h}{n} - 0.1\frac{p}{h} - 0.12\frac{p}{n}$, but $0.8 < C < 2.0$, h is the height of the construction, n is the side of the cross section normal to the wind and p is the side parallel to the wind.

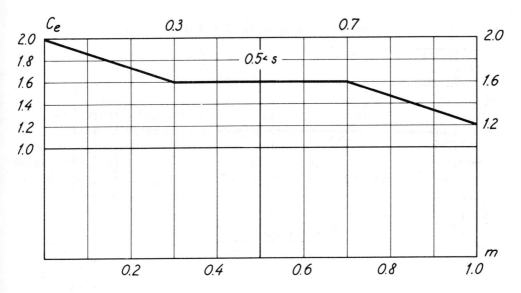

Fig. 1

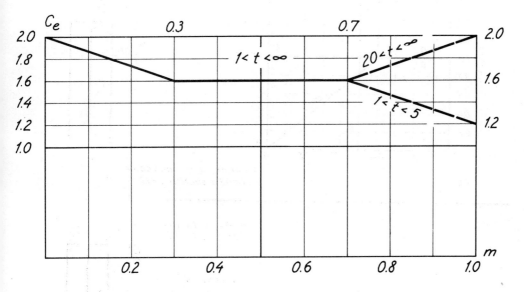

Fig. 2

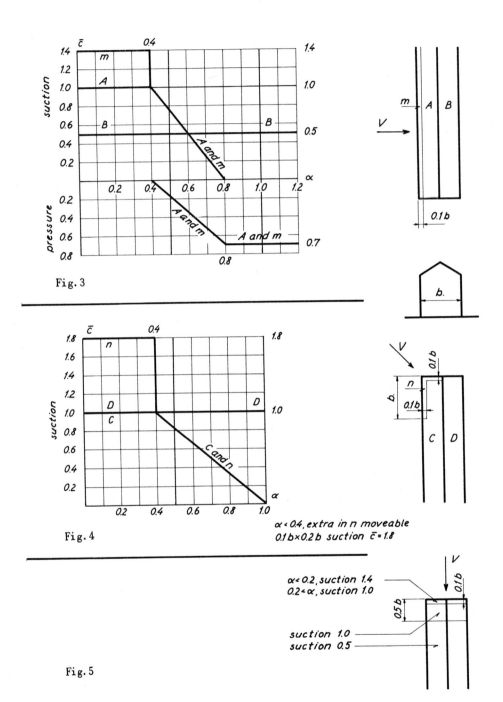

Fig. 3

Fig. 4

$\alpha < 0.4$, extra in n moveable
$0.1b \times 0.2b$ suction $\bar{c} = 1.8$

$\alpha < 0.2$, suction 1.4
$0.2 < \alpha$, suction 1.0

suction 1.0
suction 0.5

Fig. 5

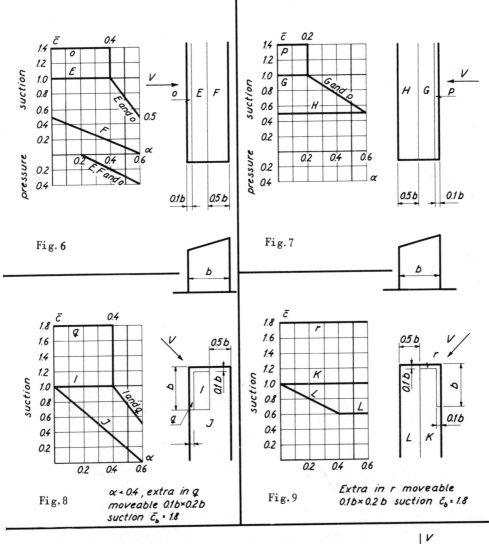

Fig. 6

Fig. 7

Fig. 8 $\alpha < 0.4$, extra in q moveable $0.1b \times 0.2b$ suction $\bar{c}_b = 1.8$

Fig. 9   Extra in r moveable $0.1b \times 0.2b$ suction $\bar{c}_b = 1.8$

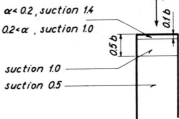

Fig. 10

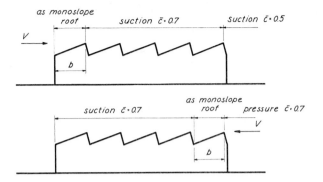

Fig. 11

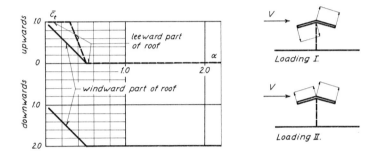

I applies to all values of α.
For $\alpha < 0.5$ also II.

Fig. 12

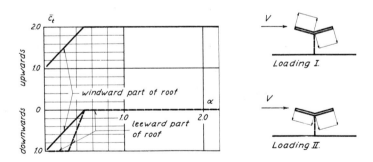

I applies to all values of α.
For $\alpha < 0.5$ also II.

Fig. 13

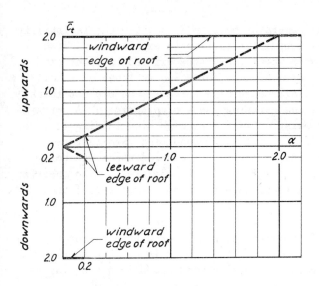
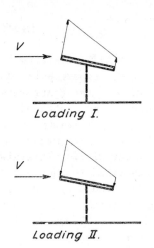

*I applies to all values of α.
For $\alpha < 0.2$ also II.*

Fig. 14

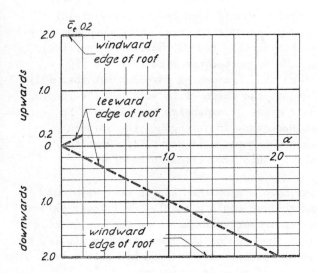
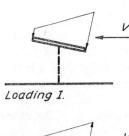
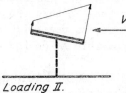

*I applies to all values of α.
For $\alpha < 0.2$ also II.*

Fig. 15

DISCUSSION ON PAPERS 19 AND 14

DR. BLAIR, referring to the papers of Session 3, doubted the reliability of model experiments in wind tunnels for full-scale experiment. With regard to the controversy as to whether a vortex existed in front of a building he instanced the pattern of snow formation in front of a building as evidence in support of the existence of a vortex. He did not think that there was much wrong in the existing codes of practice. A certain element of risk must always exist. If a very high wind loading is to be prescribed to take account of the worst wind condition then factors of safety could be reduced to unity. Not enough consideration was being given to the types of building on which the wind operates. The inertia of the building, and its dimensions, come into the assessment of the relevant gust duration.

M. d'HAVE introduced two factors which were incorporated in the Belgian code. The first was the "reduction factor" which takes into consideration the fact that at any moment the design wind speed is not acting on all parts of the building since the wind is not steady over the whole building and is not coming from the same direction. Values for the "reduction factor" are suggested in the Belgian code, but the research of Mr. Newberry should contribute to the assessment of a more realistic value. The work of Dr. Pris and Mr. Harris should lead to similarly useful results, especially since their programmes appeared to be complementary in that Dr. Pris is studying the horizontal, and Mr. Harris the vertical, correlations of wind speeds. He supported the view that too high a surface wind speed should not be used for design purposes but that full data should be available to enable the engineer to work out the loadings required. To obtain this data might be expensive but it would be less expensive than the investment necessary to strengthen the structure in order to deal with every rare occurrence. It might, for instance, be possible to design a valve which would vent excess pressures.

PROFESSOR PAGE deplored any complacency that all is well with our present wind loading code and that we could possibly adopt lower values. One of the important advances in recent years has been the realisation of the importance of the forces at the edge of the roof, and his studies of the detail of the failures in Sheffield had convinced him that a very great deal of reliance can be placed on wind tunnel tests. If the results of wind tunnel tests are compared with the full-scale pattern of failure the correlation is remarkably good. One of the interesting features of the proposed revision of the Danish code was the treatment of the "edges" of the building. In the Sheffield disaster failures occurred along the windward edges and these demonstrated that attention to detail at the edges is extremely important. The problem of chimneys is perhaps difficult to study on the scale of a wind tunnel model but in Sheffield extensive damage occurred in the lee of chimneys. This was a crucial feature and led to the suggestion that an area around a chimney breast (or a bell tower) should be marked off as an area of increased suction in the same way as the Danish code denoted the extra loading at the edges. Another common failing was for gabled walls to be pulled out, and such walls should receive detailed attention in the codes of practice.

MR. NEWBERRY wished to remove any impression that structures designed to the existing (British) codes of practice were immune from failure due to wind. There were, unfortunately, many examples which refuted this suggestion. He instanced the destruction of a school building by wind in which the roof cladding was removed with the trusses from the stanchions. The strength of this construction did in fact exceed the code requirement by some 5 percent. Mr. Newberry mentioned that he, with Mr. Scruton, had given some consideration to the revision of the existing British code and had been thinking along slightly different lines from those of this code or of the codes which had been discussed at this Conference. The approach of Mr. Scruton and himself was to treat the wind loading for code purposes in two ways, firstly the consideration of the overall forces acting on the building or structure (such as the drag forces which are the important ones from the standpoint of the main structure) and secondly to give separate consideration to the effects on the cladding. These two requirements must be treated differently because, among other considerations, of the necessity to match the gust duration with the response. For the purpose of the code, Mr. Scruton and he had suggested that the worst wind loading should be quoted because this is what the designer needs to use. For overall stability of a building this will occur with the wind directed normally to the major face. For cladding, however, it is necessary to consider the wind incident from all directions and to select the worst combination of external and internal pressures. This treatment, which we believe is of the greatest use to the practising engineer, is described in a recently published paper.*

MR. ENTWISTLE said that the importance of the problem was illustrated by the fact that in the Sheffield gales no less than 60 percent of the

---

* C. SCRUTON and C. W. NEWBERRY On the Estimation of Wind Loads for Building and Structural Design *Paper 6654 Proc. Instn. civ. Engrs. 1963, 25,* 97.

ordinary private houses were damaged. He could see no alternative to wind tunnel tests to obtain the basic data required for the codes, and suggested that the wind tunnel scientists should receive more co-operation from practising engineers as to what fields to explore. With regard to the question of whether or not a vortex is formed in front of a building, he felt that it was common experience that a door on the windward side tended to be blown in, and this indicated a positive pressure build-up. The buildings which best withstood the Sheffield gales were the heavier types with a greater margin of super load and dead load which, except for the cladding, gave them a margin against wind. Such structures should be treated differently from the present method and from other buildings with lighter cladding. The Danish code is of help here because it is possible to vary the load factor in accordance with the combination of loads, and there is a better chance of getting away with higher pressures when other loads come into consideration.

MR. CARPENTER'S concern was with the design of crane structures, in particular the type with tall mast and a revolving superstructure with a long jib. When the height of this type of crane exceeds 100 feet its stability and strength are dictated entirely by the wind. We have no guidance for considering the effect of wind directed along a long structural member such as the jib. If one member is behind the other the actual area in silhouette is small but nevertheless there is a considerable wind drag, even if the jib is allowed to run free in storm winds so that it can act as a weathervane. In such a condition the jib tends to rotate on its axis and present more or less of its side area to the wind. He would like to know how to assess the wind loads in this condition.

M. ESQUILLAN (in reply) said that every contingency could not be covered by a code. Design calculations were, of course, very good and necessary, but the builder must know the rules of his trade. Some of the points made were not new. In mountainous country the peasants know that they have to put heavy stones on the edges of their roofs to avoid losing them. Thus in many cases the norms are not met. With regard to the Sheffield disaster he wished to know how many years elapsed between occurrences of this severity. It is a question of risk and probability, and also of economics; of whether all the money available is to be spent on attempts to achieve complete safety in building construction or on some other project such as atomic research!

THE CHAIRMAN. We should remind ourselves that in this world we cannot have 100 percent safety. This fact is often overlooked.

DR. JENSEN (in reply) thought that Mr. Carpenter's crane problem could be calculated according to the Danish code of practice. With regard to the rotation problem, some mechanical clamping might be of help. In the prediction of storm wind speeds he preferred to use "yearly probabilities" rather than "return periods". The one is the reciprocal of the other but by using yearly probability one avoids anticipating conditions 100 or 500 years ahead! He questioned the use of the word "turbulence" by some of the authors; the disturbance of a stalk of grass will produce turbulence in the airflow. On the other hand, a meteorological "low" will pass South Norway in, for example, 10 hours, creating a storm over Denmark for the same time. This latter phenomenon is part of the wind climate and not due to turbulence. He considered turbulence as disturbances in the air flow created by the roughness in the terrain whereas gusts and larger disturbances were probably generated at higher levels of the atmosphere and should be treated as part of the wind climate.

PAPER 9

THE BUFFETING OF STRUCTURES
BY GUSTS

by

A. G. DAVENPORT
(University of Western Ontario, Canada)

THE BUFFETING OF STRUCTURES BY GUSTS

by

A. G. DAVENPORT

(The University of Western Ontario, Canada)

SUMMARY

THE paper attempts to trace the evolution of a satisfactory approach to the loading of structures by gusts. It is suggested that a statistical approach based on the concepts of the stationary random series appears to offer a promising solution. Some experiments to determine the aerodynamic response of structures to fluctuating turbulent flow are described. Examples are given of the application of the statistical approach to estimate the wind loading on a variety of structures, including long-span cables, suspension bridges, towers and skyscrapers.

1. STATEMENT OF THE PROBLEM

Fig.1 shows a photograph of a large 12 foot square board mounted with its centre 12 feet above ground in an open field. The board is hinged along the lower edge and reaction is taken at the centre of the board by a strain-gauge dynamometer. The board is relatively stiff. To windward of the board and outside of the disturbed airflow is a fast responding anemometer mounted at the same height as the board. With this simple apparatus it has been possible to make measurements of the force of the wind on a structure and the wind character in its general neighbourhood.

Fig.2a indicated the actual force measured on the board during a minute when a strong gust occurred. *Fig.2b,c,d* and *e* show how four structures having different but typical dynamic characteristics might respond to this short run of wind. *Fig.2b* and *c* show the response of a structure having a relatively low natural frequency (0.2 cycles per second) combined with first, light damping characteristics (1 percent critical) and second, relatively heavy damping, (20 percent critical). In *Figs.2d* and *e* a structure having a higher natural frequency (1 cycle per second) is combined with the same damping parameters.

The peak responses of the structures (in terms of the equivalent static force) are seen to be as follows:

	Units (arbitrary scale)	Percent Actual Peak Force
Actual Peak Force	600	100
Peak Equivalent static force:		
Low Frequency: light damping	770	128
" " : heavy damping	590	98
Higher Frequency: light damping	1180	197
" " : heavy damping	645	107

The peak force calculated from the peak velocity of the anemometer during the same minute was approximately 470 units.

For the present no general inferences should be drawn from the above figures since in the next minute the relative magnitudes of the peak responses could have been altogether different. The purpose in providing these figures is to show with the aid of actual measured windloads that the effects of a gusty wind on different structures can vary considerably and cannot be determined - except in the crudest terms - simply from a single instantaneous velocity. The problem of gust loading is to determine what factors affect the response to gusts and how they can be estimated.

Fig.2 has indicated two parameters affecting the susceptibility of structures to gust loading, namely
    (1)   THE NATURAL FREQUENCY
    (2)   THE DAMPING.

There are other factors which cause significant differences, these include
    (3)   THE SIZE
    (4)   THE SHAPE
    (5)   INTENSITY OF GUSTINESS and
    (6)   HEIGHT ABOVE GROUND.

Aerodynamic shape (as distinct from size) distinguishes between sharp edged or rounded structures, solid or latticed structures, and determines whether they respond to vertical and lateral gusts as well as longitudinal. The properties of the gustiness itself have been shown elsewhere[3] to be governed principally by the roughness of the ground; it is much greater in a city than over open country. It also varies with height above ground.

2. HISTORICAL BACKGROUND

While man's awareness and qualitative knowledge of the wind date back to the days before the sailing boat, quantitative estimates of its force are comparatively recent. Rouse's results presented to the Royal Society by Smeaton in 1759 are probably amongst the earliest recorded. The systematic study of wind loading can conveniently be dated from about the time of the Tay Bridge disaster in 1879. This tragedy impressed itself very forcibly on the engineering profession. The designer of the bridge, Sir Thomas Bouch was discredited and work begun on his design for a bridge across the Firth of Forth was discontinued. The responsibility was inherited by John Fowler and Benjamin Baker.

Baker's contribution to the understanding of the real action of the wind is quite considerable particularly in relation to gust action. In describing experiments he performed on large and small wind gauge boards, 300 and 1½ sq. ft. in area respectively, he stated in 1884:

"From the records generally, and from my own watching of the movements of the gauges, I have come to the conclusion that uniform velocity and pressure in a wind, whether it may prevail or not at cloud heights, can never obtain near the surface of the earth or in the neighbourhood of any bridge or other structure capable of causing eddies. Unsteady motion must be the rule in air as in water, and the threads of the currents moving at the highest velocity will strike an obstruction successively rather than simultaneously, so that the mean pressure per square foot on a large area must be less than that on a small surface from that cause alone, irrespective of possible differences in the partial vacuum at the back of the planes."

In fact, Baker found the pressures on the large board almost 1½ times that on the smaller.

Baker clearly appreciated both the spatial and transient nature of gusts. He did not appear to consider the possibility of dynamic amplification due to sequences of gusts - but this is not surprising since many of the structures of his generation possessed a rigidity which practically precluded the possibility of dynamic excitation.

Baker used the reduction in pressure over the larger area as an argument to justify using a design wind pressure of 56 lbs/sq.ft. instead of 112 lb/sq.ft. - a figure corresponding to the highest recorded wind velocity in the British Isles of 150 miles per hour. Later Sir Thomas Stanton[19] in spite of his own experimental evidence obtained in Bushy Park in support of this spatial effect, advised against its adoption as a design factor mainly on the strength of his noteworthy Tower Bridge experiments. The basis for his conclusion appear now to be somewhat dubious.[9] In the follow-up experiments at the Severn Bridge by Bailey and

Vincent, Stanton's earlier recommendation was reconfirmed but seemingly for negative rather than positive reasons.

Part of the difficulty facing experimenters at this time (1920's and 30's) was the lack of a suitable theoretical frame work into which to fit their observations. Although G.I. Taylor's statistical theory of turbulence had then largely been formulated[20] its significance was still not fully appreciated even by meteorologists. In a sense much of the work appeared to be directed towards distinguishing between possible and impossible events instead of between the probable and the improbable.

For engineers, a great deal of light was shed on the subject by a series of monumental experiments conducted during the 1930's by Sherlock at the University of Michigan[16,17,18]. From detailed wind measurements made by instruments spaced at 50 ft. intervals across a 700 ft. horizontal front and at 25 ft. intervals on a 250 ft. mast Sherlock determined gust factors for gusts of various durations and also their spatial distribution. More than anything these experiments drew the attention of engineers to the statistical nature of turbulence, its spatial properties and the time needed to build up gust pressures on a structure. The limitation of the approach lay in the fact that it did not make allowance for the possibility of sequences of gusts striking a structure nor of the probability of such an event occurring.

Somewhat earlier, in the early 1930's, Giblett and Durst had made a pioneer study of the structure of the wind at the Royal Airship Works, Cardington. In these experiments, in spite of the enormous computational task, statistical methods on the lines of G.I. Taylor's theory were applied almost for the first time to determine the space and time correlations of the gustiness. This line of investigation was in many ways well ahead of its time and it seems probable that had airships not come to the sudden end they did and this work had been allowed to continue, that the understanding of wind loading would have been advanced considerably.

It is interesting to note that with the aid of modern computing techniques, it has been possible to reanalyse the high quality measurements obtained by both Sherlock and Giblett and fit them into the framework of the statistical theory of turbulence[6].

It is apparent that at this time most of the thoughts on wind loading were that it was of a quasi-static nature, and that the highest instantaneous pressures were what mattered. The importance of the load history did not seem to weigh heavily. The lashing of trees in the wind, it seems, was not linked with the behaviour of engineering structures! One of the first papers to discuss this problem was by Horne[12] who determined the duration of a "sharp-edged" gust necessary to collapse a simple portal frame. Although a highly idealized representation of a gust, it was an interesting approach to the problem and in some ways was analogous to the sharp-edged gust approach used in aircraft circles.

A very notable advance in the approach to aircraft gust loading was presented in a paper by H.W. Liepmann[13] entitled "The application of statistical concepts to the buffeting problem". In this he points out the futility of trying to define gusts explicitly. Instead it was necessary to treat the problem statistically since it was only in these terms that turbulence itself could be defined. The approach was based on the statistical concepts of the stationary random process. As applied to aircraft gust loading, the procedure followed is illustrated in *Fig.3*. First the gust fluctuations are broken down frequency by frequency into their spectrum by harmonic analysis. Knowing the response of aircraft wings to sinusoidal gusts of different wavelength it was then possible to determine the corresponding spectrum of lift forces. Knowing the dynamic amplification of the aircraft, that is the response to sinusoidal lift forces, in turn, the response (stress, deflexion, etc.) spectrum of the aircraft could be determined. The area under the response spectrum is a measure of the mean square response of the aircraft, or in other words, the variance of the probability distribution of the response. Since turbulence generally followed a more or less Gaussian distribution the response tended to as well. In this case the mean and variance are sufficient to define the probability distribution completely.

This approach to gust loading was a natural outcome of two earlier developments: first the great success of Taylor's statistical approach in defining turbulence itself, and second, the success and development of similar approaches in the response of electronic filters to random noise in the communications field. The method seemed obviously suitable for the gust loading of civil engineering structures and the writer[7] suggested this in a paper entitled "The application of statistical concepts to the wind loading of structures."

In more recent papers,[4,5] the application of the method has been extended from the original application to a simple one-degree-of-freedom point structure, to more complicated, spatially extended, multi-degree of freedom structures, capable of being excited in several modes of vibration, not only longitudinally but also laterally and torsionally.

3. THE STATISTICAL APPROACH

3.1. Wind Properties

In determining the effect of wind on a structure the gust loading question might arise in the following way. A designer wishes to determine the maximum stress likely to be exceeded in a structure only once in say

50 years, or 200 years or possibly he wishes to know the complete probability distribution of yearly maximum stresses. The reasoning might follow these lines. It was shown in a previous paper,[3] that the mean gradient wind speed having a return period of r years (in this case 50 or 200 years) or a probability of $(1-\frac{1}{r})$ was

$$V_G(r) = V_G - \frac{1}{a_G} \log_e r$$

V_G and $\frac{1}{a_G}$ are the mode and dispersion factor for the extreme gradient wind speed field at the site of the structure. This information can be used to determine the corresponding mean wind speed at some specific height above ground provided that some estimate of the ground roughness and exposure of the site is available. This can be obtained by inspection or preferably by site investigation. (The ground roughness is the principle parameter governing the mean wind speed profile and the gustiness). Knowing the mean wind speed at structure height corresponding to the gradient speed $V_G(r)$ and the ground roughness, the vertical and horizontal gust spectra can be defined. It is upon these that estimates of the gust forces superimposed on the mean wind forces must be based.

3.2. Aerodynamic Forces

Turning now to the aerodynamic pressures this flow produces on a structure, some insight can be obtained if, instead of the whole spectrum or gusts, we suppose that the airflow consists of the mean flow $\bar{V}$ with a superimposed gust fluctuation $v\sin 2\pi nt$ which is moderately small compared to $\bar{V}$. The simplest model of the response is one in which the aerodynamic force consists of a mean force $\bar{P}$ with a superimposed fluctuation $p\sin 2\pi nt$. If this response was quasi-steady then P would be $\frac{1}{2}\rho C\bar{V}^2$ where C is the aerodynamic force coefficient and the value of p would be $\rho C\bar{V}v$ (neglecting v^2 terms).

This quasi-steady response is of course likely to obtain when the gusts are extremely large compared to the size of the structure and slowly varying: - the important parameter determining this is $\frac{D}{\lambda}$ - the ratio of the structural diameter to the gust wavelength λ, where $\lambda = \frac{V}{n}$. When the gusts are large, $\frac{D}{\lambda}$ is small. In the range where the gusts wavelength and the structure are more nearly equal (say $0.01 < \frac{D}{\lambda} < 100$) a quasi-steady response is not likely to result.

There are several reasons for this: as discussed in another paper,[3] the effective size of a gust is determined by its scale (or semi-scale) which is approximately proportional to the wavelength. For example, the lateral semi-scale of a longitudinal gust was found to be roughly 1/30th of the wavelength and the vertical semi-scale about 1/8th of the wavelength. This means that smaller wavelength gusts are more localized and therefore less effective in producing pressures than larger gusts which can completely engulf a structure. Eventually, a stage must be reached when the gust wavelength is so small compared to the size of the structure that its effect is no longer perceptible.

Another variable factor concerns the value of the force coefficient C. In unsteady flow, this is likely to be different to the steady flow value. This is because in unsteady flow, forces are incurred in decelerating and accelerating fluid (the virtual mass effect) which are in phase with the acceleration: there also appears to be a definite tendency for the drag coefficient to rise. This is indicated in *Fig.4a* and *b*, which shows the fluctuating drag coefficient of a triangular lattice truss mast section with a solidity of roughly 40%, when placed in a uniform fluctuating flow (as opposed to a turbulent fluctuating flow). The uniform fluctuations in the relative flow were obtained by oscillating the model on the end of a pendulum immersed in a uniform steady flow.[8]

The rise in effective drag appears to be noticeable for wavelengths between 10 and 1 diameter. This result, it should be emphasized, appertains only to the fluctuating component of the drag; the steady component was still governed by the normal steady flow drag coefficient. Other results for the flat plate showing a similar trend were reported previously.[7]

In a turbulent flow, two opposing tendencies therefore seem to be at work determining the effective aerodynamic force coefficient linking the velocity fluctuations at a point to the force fluctuations. The tendency for the force coefficient to rise with increase in frequency is offset by the decrease in the spatial correlations. Eventually, it seems the effect of the decreasing spatial correlation must swamp whatever increase may be due to increases of C. The exact way in which this happens is not as yet clearly understood, although most important.

Instead of dealing with individual harmonic components of flow fluctuation, we now turn to the complete spectra of the velocity and pressure. Suppose that the normalized velocity spectrum is $\dfrac{n\,S_v(n)}{\bar{V}^2}$ and the normalized pressure spectrum is $\dfrac{n\,S_p(n)}{\bar{p}^2}$ and that

$$\frac{n\,S_p(n)}{\bar{p}^2} = |\chi(n)|^2 \; 4 \; \frac{n\,S_v(n)}{\bar{V}^2}$$

where $|\chi(n)|^2$ is the "aerodynamic admittance" and is the ratio of the square of a fluctuating-flow aerodynamic force coefficient to the mean flow force coefficient. The quantity 4 appears in the equation so that the aerodynamic admittance, $|\chi(n)|^2$ is approximately unity for quasi-steady flow: at very high frequencies, the spatial correlation requirements suggest that $|\chi(n)|^2$ tends to zero.

Recently, the author has been engaged with two sets of experiments designed to unravel the rudimentary nature of the aerodynamic admittance for simple bluff shapes. One set of simple experiments was carried out at the University of Western Ontario on the large 12-foot square board situated out-of-doors and illustrated in *Fig.1*: the other set of experiments was conducted with R.L. Wardlaw at the Canadian National Aeronautical Establishment. Here, the turbulent flow was artificially induced in a wind tunnel and the models were much smaller, having a nominal dimension of the order of an inch or so, and consisted of rectangular discs and panels of an infinite flat plate. Drag forces on the discs placed normal to the flow were measured by a small dynamometer with a sensitivity of about one-millionth part of a pound and a flat frequency response up to about 300-500 cycles per second.

As examples, the normalized force spectrum on a 1½" square disc and the normalized velocity spectrum (obtained using a hot-wire anemometer) are shown in *Fig.5*. The ratio of the two spectra gives the effective aerodynamic admittance $|\chi(n)|^2$; values of which, for the square and rectangular discs, are given in *Fig.6*. Unfortunately, the frequency range of investigation does not extend down far enough to the point where the response is quasi-static and the two spectra merge. The results indicate strongly the decrease in effectiveness of gusts as wavelength gets smaller, i.e. increasing reduced frequency; they also suggest that the aerodynamic admittance for a longer disc is less, implying that the average gust pressure is less. This fact is more strongly indicated in *Fig.7* by the measurements of aerodynamic admittance for the panels of an infinite flat plate (rectangular discs flanked by long dummy plates). Here the trend is obviously 'the larger the area, the smaller the effective gust pressure.'

Measurements on the large board of the same quantities - the spectra of force and of the velocity (measured a sufficient distance to windward of the board to be undisturbed by it) - are shown in *Fig.8*. In addition, measurements of the mean load and the mean velocity were made. For the three runs recorded, the mean velocity, the mean load and the mean load computed from $\bar{V}^2$ the mean velocity squared and using the steady flow drag coefficient of 1.15 are as follows:

Run:	Mean Velocity (ft/sec)	Mean Load (lb)	Computed Mean Load (lb)
Dec. 5	20.2	88.2	87.8
Jan. 7[2]	26.5	136	139

The good agreement between the measured and computed mean load tends to confirm that in spite of the high intensity of turbulence (about 25%), the mean load is still about the same as it would be in steady flow.

The spectra of the fluctuating force and velocity do not at first sight appear to show the expected trend. Instead of the force spectrum decreasing more rapidly than the velocity spectrum, giving an aerodynamic admittance which decreases with frequency as in the wind tunnel tests, the two spectra in *Fig.8c* for example, appear to decrease at much the same rate. There are three factors contributing to this apparent inconsistency. The first relates to the method of support of the board which was pivoted at the lower edge with the reaction measured at the centre of the board. It can be shown that, although this leads to quite satisfactory estimates of the mean drag of the board and of the effects of large, well-correlated gusts, it tends to over-estimate the mean square drag for smaller gusts which are likely to be only locally correlated, particularly those striking the board near the top. Second, due to the proximity to the ground the airflow will tend to be drawn over the top of the board rather than symmetrically around it, thus tending to reinforce those forces which produce a great reaction at the centre. Third, although the anemometer is extremely light and fast responding, it nevertheless slightly under-estimates the high frequency end of the velocity spectrum. All these factors will tend to reduce the convergence of the spectra at the higher frequencies. If allowance was made for them the spectra of *Fig.5* and *Fig.8* would be more comparable.

These board experiments have already served a useful purpose however; they have first shown the mean force to be in agreement with measurements of the mean velocity as mentioned above. Second, they indicate that the ratio of the force spectrum of the velocity spectrum at lower frequencies is roughly equal to 4 (see *Fig.8c*) - the value which a quasi-steady approach would lead us to expect. The behaviour at higher frequencies still requires further investigation.

For latticed structures in which the drag is generated locally by a large framework of separate small members, the admittance can probably be estimated reasonably adequately by considering the average velocity correlation over the section as a whole. This approach, which was taken with the deck of a suspension bridge, is illustrated in *Fig.9*. Using a velocity scale of 1/7th of the wavelength, the average correlation of velocity over the section is plotted as a function of the reduced frequency of the section $\frac{nD}{V}$ in *Fig.10*. For trusses, this correlation is probably equivalent to an aerodynamic admittance. Interestingly enough, the values given in *Fig.10* are not far off the measured values of the aerodynamic admittance for a thin strip of the infinite flat plate given in *Fig.7*.

3.3. Structural Response

The wide variations of structural response to gusts have already been indicated in *Fig.2*. With one degree of freedom systems, estimation of the response using the spectral approach is reasonably straight forward and is indicated by *Fig.3*.

If the system is lightly damped, a large bulk of the response spectrum (almost all in some structures) will occur at or near the natural frequency. The area under this spectral peak can be shown to be approximately

$$\begin{pmatrix}\text{Mean square}\\ \text{response}\end{pmatrix} \approx \frac{\pi^2}{\delta} \times \begin{pmatrix}\text{ordinate of the force spectrum}\\ \text{at the resonant frequency.}\end{pmatrix}$$

Here δ is the logarithmic damping decrement ($= 2\pi \times$ critical damping ratio). The damping consists of mechanical damping and of aerodynamic damping. The aerodynamic drag damping, for example, is of the form

$$\delta = \frac{\bar{P}}{n_o \bar{V} m}$$

$\bar{P}$ being the mean drag, n_o the natural frequency, and m the mass. The area under the spectrum gives the mean square fluctuation. For most structures, the peak response works out to be roughly 3.5-4.0 times the R.M.S. fluctuation in excess of the response to the mean (hourly) wind.

For multi-degree of freedom structures, the approach is a little more involved. Basically, it seems, the response should be estimated in each mode separately and then the results super-imposed. Without going into detail, it can be shown that the spectrum of the response (deflexion, bending moment, etc.) in each mode consists of the product of four terms: (a) the gust velocity spectrum, (b) the aerodynamic admittance, (c) the mechanical admittance, and (d) the joint acceptance. The last determines the correlation between the gust load distribution and the mode shape. The total area under this spectrum represents the variance of the response. If a structure is (a) lightly damped, so that the damping is predominantly aerodynamic, (b) very long span, so that the span is great compared to the lateral spanwise scale of the gust at the natural frequency, then it can be shown that the variance is of the form:

$$\text{constant} \times \frac{\bar{V}}{n_r \ell} \times \frac{m}{\bar{P}} \times n_r \bar{V} \left| \chi(n_r) \right|^2 \times \frac{n_r S_v(n_r)}{\bar{V}^2}$$

Here $\frac{\bar{V}}{n_r \ell}$ represents the ratio of the wavelength at the natural frequency to the span, $\frac{m}{\bar{P}}$ represents the mass of the structure per unit drag, $\left|\chi(n)\right|^2$ is the aerodynamic admittance and $\dfrac{nr\ S_v(nr)}{\overline{V}^2}$ the normalized gust spectrum.

The basis for this expression has been indicated earlier.[5] Although the application of this particular expression is restricted, several features emerge about gust response which help in a general understanding. It indicates first that the gust response is generally less if the natural frequency is higher, since resonance will then coincide with higher frequency gusts containing less energy and which are less well correlated. Furthermore, the aerodynamic admittance will usually be smaller. Second, the response will be smaller for longer structures due to the smaller spanwise correlation. And, third, the response will be smaller for structures having a high ratio of drag to weight, since this will result in higher aerodynamic damping.

3.4. Fatigue

The very real problem of fatigue that could be caused by gust loading, is suggested by *Fig.2c*. In fact, fatigue by wind of light structures such as lamp standards, and signs is not uncommon. The problem is difficult to resolve quantitatively except in broad terms. Knowledge of the response spectrum does, however, permit a count to be made of the number of occurrences of given stress levels per unit time.[6] One of the unknowns, however, is the number of hours per year for which given wind strengths prevail.

4. APPLICATIONS TO PARTICULAR STRUCTURES

In the following sections, the above remarks are amplified by discussing briefly the response of several structures which have been analysed elsewhere.

4.1. Water Tower and Arc Lamp Standard

The wind loading of two structures of these types was analysed according to the statistical methods in a paper published in 1961. The two structures are shown in *Fig.11*. They were purposely chosen first because in practice they would normally be designed to resist almost precisely the same wind load (each had the same frontal area) and second, because their natural frequencies would be contrastingly different (see *Fig.11*). On the one hand the natural frequency of the arc-lamp was low enough (0.5 cycles/sec) to make the reduced frequency at resonance $\dfrac{nD}{\overline{V}} < 0.1$ at the design wind velocity, implying that gusts would be almost fully correlated over the structure. On the other hand the water tower was supposed to be rigid with a high natural frequency of about 5 cycles/second. This meant a reduced frequency at resonance closer to 1, at which it was assumed the gusts would be very poorly correlated and almost ineffective. For purposes of illustration the peak once-in-fifty year equivalent static wind loads were worked out for both structures, for both open country and city conditions in a region where the once-in-fifty year gradient wind speed was about 110 mi/hr. The results were as follows.

Equivalent Static Wind Load - lb/ft^2

	Mean Hourly Wind Load	Gust Load	Total Load	Ratio:	$\frac{\text{Gust Load}}{\text{Mean Load}}$
Open Country					
Arc Lamp	15.3	45.7	61		3.0
Water Tower	15.3	19.7	35		1.3
City					
Arc Lamp	3.3	23.7	27		7.2
Water Tower	3.3	10.7	14		3.3

In current practice all of the above structures would have been designed for about the same wind load (probably about 30 lb/sq. ft.). Had once in 200 year wind loads been considered more appropriate the above figures would have to be increased by 20%.

It should be noted that in all these structures the gust loading is a larger proportion of the total load than the mean wind loading. This means that more than half the load is of a dynamic character.

4.2. Long-span Cable

This structure served to illustrate the application of statistical concepts to a line-like structure[4]. A cable span of 4000 ft. with a dip of 270 ft. was considered. The natural periods were very long i.e. very low natural frequencies - being 15 and 8 seconds for the first two horizontal modes and 8.25 and 5.8 for the vertical. On account of the low natural frequencies and the relatively small diameter of the structure the reduced frequencies at resonance with the several modes were all much less than 0.1 and the aerodynamics could therefore be taken as for quasi-steady conditions. The site was taken as being moderately rough ($K = 0.01$) and the mean wind velocity 100 ft/sec.

The drag force per unit cable weight was taken as $3.3 \times 10^{-5} V^2$ (V is velocity in ft/sec). The response in the various modes of the structure were found to be as follows:

Mode		Period (seconds)	Mode	Peak Deflexion (feet)
Horizontal	1st.	15	$\sin \pi \frac{x}{\ell}$	47 (15)
	2nd.	8.00	$\sin 2\pi \frac{x}{\ell}$	12 (4)
Vertical	1st.	8.25	$\sin 2\pi \frac{x}{\ell}$	5 (2)
	2nd.	5.84	$\sin 3\pi \frac{x}{\ell}$	2 (1)

By comparison, the mean deflexion at the mid-point is approximately 90 ft. In the original calculation, the lateral scale of the longitudinal and vertical gusts was taken at roughly 1/7 of the wavelength. This corresponds to unstable atmosphere conditions. It now seems that this assumption is somewhat conservative and a more realistic value (suggested elsewhere,[3] for the lateral scale might be 1/30 of the wavelength. In this case, the peak gust deflexions will be reduced by approximately 1/2 (i.e. $\sqrt{\frac{7}{30}}$) and have the values in the brackets.

For this structure, the gust load is only about 15 per cent of the mean load, the spatial reduction of gust pressures and high aerodynamic damping are the principal reasons.

4.3. Suspension Bridge

The suspension bridge provides an interesting contrast to the cable span. The results of an analysis[5] of the response of a 3,300 ft. span suspension bridge to an 85 mi./hr. mean wind are shown in *Fig.12*. These results were also obtained using the somewhat conservative value for the lateral scale of the gusts - 1/7th of the wavelength; using the smaller lateral scale used in the cable span example (1/30th of the wavelength) - reduces the gust effects shown by 50%. With this reduction the ratio of maximum shear force contributions of gust and mean wind is about 1.1, and the ratio of maximum bending moment contributions about 1.4. The reason for the greater susceptibility of the suspension bridge to gusts compared to the cable span lies mainly in the smaller aerodynamic drag damping of the former. This is principally governed by the drag force per unit weight of structure; for the cable span, this is roughly $3.3 \times 10^{-5} V^2$, for the suspension bridge about $0.3 \times 10^{-5} V^2$. The aerodynamic damping of the suspension bridge to therefore about $\frac{1}{10}$ th that of the cable.

Bearing in mind the much smaller stiffness of the deck for bending in the vertical plane, the effect of vertical gusts are seen to be just as significant as the horizontal gusts.

The results were based on a surface drag coefficient of 0.01 (a roughness between woodland and open grassland). If the fetch of the prevailing wind was predominantly over open water, then the drag coefficient might be as low as 0.001 and the intensity of gustiness and hence the ratio of gust to mean loading reduced by a factor of $\sqrt{10}$. On the other hand, the more open exposure would result in higher mean wind velocities.

In analysing this structure, it was found that it was only the response in the first two or three lateral modes and the first three or four vertical modes that mattered.

4.4. Tall Masts

Apart from their own dead load, practically the only loads these structures have to withstand are due to wind. Determination of the wind effect is therefore a primary consideration.

The results of an analysis of a typical guyed mast 500 feet high, guyed at three levels with a cantilevered portion at the top, are shown in Fig.13.[8] In this illustration, comparison is made between the response of a mast situated in open-country with the response of one in a city nearby, both masts being subject to the same gradient wind speed.

An important fact that emerges is that the overall response in a city is significantly less than it is in open country; the overall maximum shear force being approximately 1/3 of that in open country and the bending moment 1/2.5 of that in open country. A further contrast that can be made is between the ratio of maximum gust loading to the maximum mean wind loading for the city and open country conditions. For the shear force (S.F.) and bending moment (B.M.), the ratios are as follows:

	City	Open Country
$\dfrac{\text{S.F. due to Gusts}}{\text{S.F. due to mean wind}}$	0.5	0.8
$\dfrac{\text{B.M. due to Gusts}}{\text{B.M. due to mean wind}}$	0.9	0.5

To some extent, these figures may seem contradictory, particularly with respect to the shear force figures which indicate a higher relative gust shear force in open country than in the city. The explanation for this lies partly in design of the mast. It was designed according to conventional methods for open country conditions, in which, the cable sizes are chosen to give the mast more or less straight line deflexion under a wind velocity increasing with height more or less according to the seventh power law; that is, a load increasing with height according to $Z^{2/7}$ - in fact $Z^{0.3}$. Under city conditions, the increase of wind velocity with height is much more rapid and the wind load increases approximately as $Z^{0.8}$. Under this loading, the same mast has a deflexion far from a straight line and a great deal of the unbalanced loading is taken to the ground and lower guys directly by the mast, particularly in shear. Thus the mean shear forces in the city tend to be relatively higher than in the open country.

Although these figures are probably fairly indicative of the behaviour of a tall guyed mast, under wind loading, the analysis may turn out to have been based on an over-simplified model of the mast's dynamic behaviour, particularly for tall masts. The area of over-simplification lies in the assumed behaviour of guy wires. Recent experiments at the University of Western Ontario have for example, indicated that the earlier quasi-static model of their behaviour (used in obtaining the results of *Fig.13*) can be grossly in error. Under dynamic excitation, it was found that the guy wires provide substantial amounts of damping, can give a restitution modulus (the spring constant), which varies widely with frequency and which, over certain frequency ranges can be strongly negative. The last implies that, rather than supporting the mast, the guys are tending to pull it over.

Mechanical damping in these structures tends to be fairly small as can be seen from a damping decay trace of a 500 ft. mast shown in *Fig.14*. The curve indicates a logarithmic decrement of about 0.06 or a critical damping ratio of about 1 per cent.

The dynamic character of the loading can be seen from the short section of oscillograph trace obtained from an accelerometer placed on top of a 1000 foot guyed mast. It is interesting to note that Professor Tachu Naito (1959) of Waseda University in Tokyo has reported that the measured acceleration on the self-supporting Tokyo mast (roughly 1000 feet high - comparable to the Eiffel Tower) reached 0.410 that of gravity during a typhoon with winds of about 90 miles per hour.

4.5. Skyscrapers

In view of the current trend towards taller buildings the effect of wind on such structures takes on a special significance. It so happens that practically the only published observations on the behaviour of a full scale structure of any size in the wind have been made on a skyscraper - in fact the Empire State Building (observations are also currently being made on other large buildings). They are described by Rathbun in a paper entitled "Wind Forces on Tall Buildings" published in 1940. Four different types of measurement were made

(1) Manometer tube measurements of the pressure at three floors (36th, 55th, and 75th).
(2) Wind velocity measurements from the anemometer on the top of the mast head. (1250 ft.)
(3) Sway measurements between 6th and 86th floors by vertical collimator measurements (also a plumb-bob) and
(4) Extensometer measurements taken on four columns on the 28th floor.

From the manometer tube measurements and wind velocity measurements it was possible to make comparisons with the pressures measured on the 5 ft. high wind tunnel model tested by Hugh L. Dryden and G.C. Hill then of the National Bureau of Standards. From these Rathbun states:

"A comparison of the pressures on the model and those on the building shows clearly that the natural wind movements are not at all like those in a wind tunnel".

On the basis of the collimator and extensometer measurements Rathbun states:

"In effect the building is a huge cantilever and as such has two distinct movements when acted upon by a wind storm; it deflects from the vertical and it vibrates with a definite period similar to the tines of a tuning fork when struck. A steady wind causes deflexions only whereas a gusty wind will set up vibrations with amplitudes that vary with the strength and character of the storm. Under the action of a strong wind the building vibrates about a mean position in the direction parallel to its narrow side and with a different period in a perpendicular direction. Apparently, these two vibrating motions are independent of each other differing both in amplitude and period".

The amplitude of the vibratory motion appeared to be of roughly the same magnitude as the mean deflexion - although the latter was not easy to determine owing to the difficulty of estimating the true origin of the deflexion. The mean and fluctuating stresses produced would of course, be roughly in the same ratio.

A trace of the stress measured in one of the columns is given in *Fig.16*. This emphasizes the dynamic character of the wind action.

The problem of wind action on tall buildings is important from two points of view - not only the structural security but also because of the problem of vibration. Even winds too weak to cause any structural distress may be strong enough to produce feelings of insecurity in the occupants or prevent some activities such as sensitive weighing. An example of such vibration is given by the following extract from the Miami Herald of October 6th 1948 which appeared following a hurricane: (quoted by Saffir - 1957)

"County Jail prisoners high up in Dade County skyscraper courthouse were disturbed Tuesday night as the building swayed in high winds. The chief deputy reported that some of the prisoners kept track of the swaying by watching the movement of water run into washbowls."

A recent paper by Philcox entitled "Some Recent Developments in the Design of High Buildings in Hong Kong" has suggested that the problem of vibration can exist with even moderately tall buildings as indicated by the following quotation.

"Buildings with height to width ratios of less than 2 which were designed without wind prior to 1958 (which year saw the introduction of the requirement that all buildings in Hong Kong should be designed against the effect of wind irrespective of height - width ratio) have in a recent typhoon exhibited the tendency to vibrate to an uncomfortable degree with, however, no signs of structural failure. A shear wall stiffened building has on the other hand, been reported to be almost free of vibration at the height of the typhoon".

The last sentence raises the question of what factors determine whether or not a tall building will vibrate seriously. Whereas the mean deflexion and stress are determined primarily by the stiffness, the fluctuating deflexion and stress depend principally on the damping, the natural frequencies, the building size, and the modes of vibration.

While allowance is made in present day design for variations in stiffness in estimating the wind effects on structures, no allowance is made for variations in either damping or natural frequency. Safety factors and the conservative nature of the wind loadings have evidently been generally adequate to blanket even the more unfavourable combinations of these latter parameters. It is nevertheless pertinent to point out one or two trends in structural design that are likely to affect this picture: in the field of multi-storey steel frame structures two of the more significant are perhaps the substitution of welding for rivetting and the trend toward light curtain walls in place of the more traditional masonry filler wall.

The Empire State Building itself had a rivetted steel frame with masonry and concrete filler walls. From the plumb-bob measurements it was evident that there was considerable plastic action affecting the movement of the building. This could be attributed to the frictional forces induced in the rivetted joints of the steel skeleton and to the chafing of the masonry infill. During vibration these forces would provide a powerful damping action which would be absent in a modern monolithic structure. It is difficult to assess qualitatively the ratio of the damping in the two cases - the traditional and the modern - but what little evidence there is from earthquake research suggests that it might be as great as 10 (20% critical damping in the traditional and 2% in the modern). This implies that the dynamic magnification in the modern structure at the natural frequency will be about 10 times as great as in the traditional. The RMS amplitudes will be more or less in proportion to the square root of this ratio i.e. about 3.

A further factor which emerges from a comparison of traditional and modern structures is the difference in the stiffness to weight ratio, which is a useful measure of the natural frequency. Data given by Blume in connection with earthquake design would suggest that while the weight of the traditional structure is of the order of 1-1/2 to 2 times as great as for an equivalent modern structure its stiffness may be four to ten times as great (estimates for the Empire State building were about five). The reason for the greater stiffness lies of course in the masonry infill which is discounted in the design of the steel skeleton. These values in fact, imply that the period of vibration of the modern structures will be 1-1/2 to 2-1/2 times as long as for the equivalent traditional structure.

While a longer period of vibration generally improves a structure's resistance to earthquake the same would not on the whole appear to be true for wind loading. The reason is that the longer natural period then coincides with longer wave length gusts which, as we have seen, are more extensive and contain more energy. This can be seen in *Fig.17* which compares the response of two hypothetical skyscrapers of "modern" and "traditional" construction in a 100 ft./sec. wind. The modern structure is seen to have an RMS effective gust loading (superimposed on the mean load) of four or five times that on the traditional structure. This is not the whole story, however, since the "modern" structures are generally better tied together than the traditional structures which are known to suffer damage in the form of plaster and masonry cracking at comparatively low stresses in the steel frame. In addition, the response of the modern structure can be predicted more accurately and, provided realistic design assumptions are used, the structure can be tailored more closely to the loads actually imposed.

CONCLUSION:

Two main points emerge from this paper. First, that it now appears possible to approach the problem of gust loading from a fundamental rather than empirical point of view using the statistical concepts of the stationary random series. Much research is still needed to verify the assumptions used and in particular to determine the aerodynamic functions for the translation of gust velocity into pressures - but nevertheless the framework for treating the problem seems to be appropriate. The second point to emerge is the considerable variation in the susceptibility of structures to gusts due to normal variations in such structural properties as damping, natural frequency, modes of vibration, size and shape, as well as the influences of the site exposure. In the examples given, for instance, the ratio of super-imposed gust loading to mean wind loading ranged from .15 to 7.0. In another paper,[3] the large variations in mean wind loading were alluded to. The overall range of variation points to the advantages to be obtained from a detailed analysis of wind loading as opposed to the unquestioned adoption of values taken from codes and manuals.

The definition of wind loading that we arrive at is one in which the occurrence of certain wind effects (stresses, deflexions, etc.) are associated with certain probabilities (or expected return periods). That is to say, we define the probability distributions of the particular wind effects and the factors which govern these distributions. That is as far as we can go.

The question still to be answered is how this information can be rationally incorporated into a design procedure.

REFERENCES

1. BAILEY, A. and VINCENT, N. D. G., "Wind pressure experiments at the Severn Bridge". *J. Inst. civ. Engrs.*, 1939, 11, 363.
2. BAKER, B., "The Forth Bridge". *Engineering*, 1884, 38, 213.
3. DAVENPORT, A. G., "The relationship of wind structure to wind loading". *(Paper for presentation at the International Symposium on the Effects of Wind on Structures, London, England, June 1963.)*
4. DAVENPORT, A. G., "The response of slender line-like structures to a gusty wind". *Proc. Inst. civ. Engrs.*, 1962, 23, 389.
5. DAVENPORT, A. G., "Buffetting of a suspension bridge by storm winds". *Proc. Amer. Soc. civ. Engrs. J. struct. Div.*, 1962, 88, 233.
6. DAVENPORT, A. G., "The spectrum of horizontal gustiness near the ground in high winds". *Quart. J. roy. meteor. Soc.*, 1961, 87, 194.
7. DAVENPORT, A. G., "The application of statistical concepts to the wind loading of structures". *Proc. Inst. civ. Engrs.*, 1961, 19, 449.
8. DAVENPORT, A. G., (1961)c "A statistical approach to the treatment of wind loading on tall masts and suspension bridges". *Ph.D. thesis University of Bristol, England*, 1.
9. DAVENPORT, A. G., "Wind loading on Structures". *National Research Council of Canada, Division of Building Research, 1960, Technical Paper No. 88*.
10. DRYDEN, H. L. and HILL, G. C., "Wind pressure on a model of the Empire State Building". *J. Res. nat. Bur. Stand.*, 1933, 10, 493.
11. GIBLETT, M. A., "Structure of wind over level country". *Geophys. Mems. #54, 1932, Met. Office, London*.
12. HORNE, M. R., "Wind loads on structures". *J. Inst. civ. Engrs.*, 1950, 33, 155.
13. LIEPMANN, H. W., "On the application of statistical concepts to the buffeting problem". *J. aero. Sci.*, 1952, 19, 793.
14. PHILCOX, K. T., "Some recent developments in the design of high buildings in Hong Kong". *The Structural Engineer*, 1962, 40, 303.
15. RATHBUN, J. C., "Wind forces on a tall building". *Trans. Amer. Soc. civ. Engrs.*, 1940, 105, (2056) 1.
16. SHERLOCK, R. H., "Gust factors for the design of buildings". *International Assoc. for Bridge and Struc. Engineering*, 1947, 8, 204-236.
17. SHERLOCK, R. H., "Variation of wind velocity and gusts with height". *Trans. Amer. Soc. civ. Engrs.*, 1953, 118, (2553).
18. SHERLOCK, R. H. and STOUT, M. B., "Storm loading and strength of wood pole lines". *Edison Electric Inst.*, 1936.
19. STANTON, T. E., "Report on the measurement of the pressure of the wind on structures". *Proc. Inst. civ. Engrs.*, 1925, 219, 125.
20. TAYLOR, G. I., "The statistical theory of turbulence". *Proc. roy. Soc.A.*, 1935, 151, 421.

Fig.1. Photograph showing the 12 foot square board for measuring the instantaneous gust pressures in the natural wind. A fast response anemometer is mounted on the pole in the foreground.

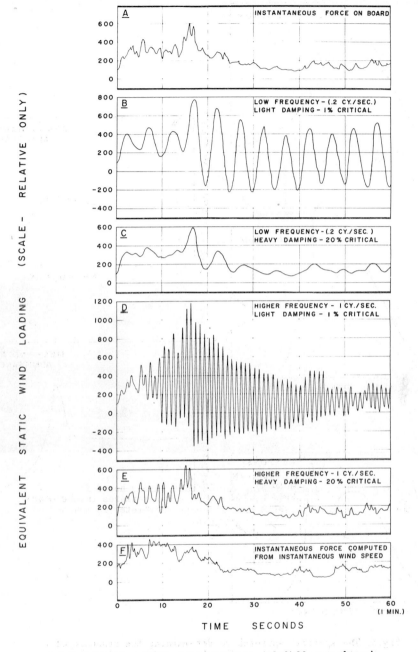

Fig. 2. Response of several structures of different dynamic characteristics to the instantaneous wind loading sustained by a 12 ft. square board.

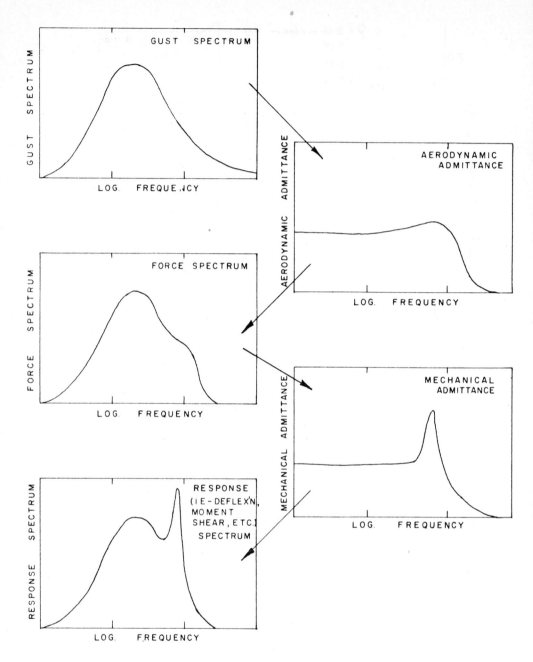

Fig.3. The spectral approach to determining the response of a structure to gust loading (area under spectrum = mean square or variance of response).

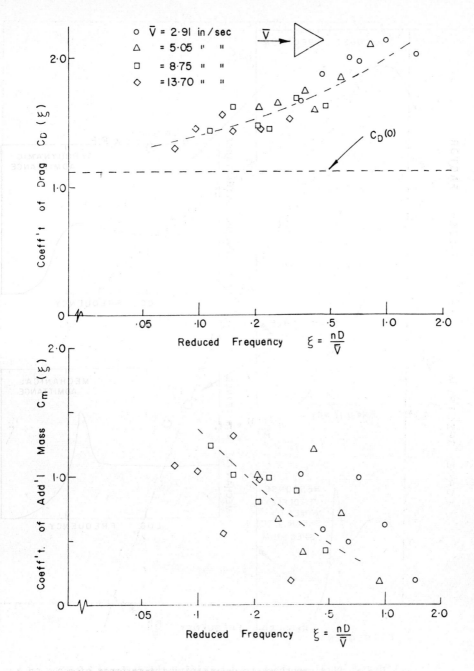

Fig. 4. Coefficients of drag and additional mass of triangular lattice truss in fluctuating flow.

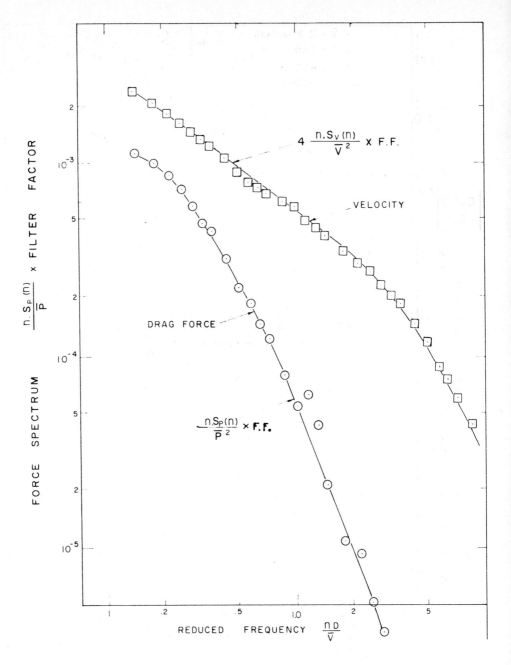

Fig.5. Wind tunnel measurements of the drag force spectrum on a 1-1/2" square disc and the wind velocity spectrum in strong turbulent flow.

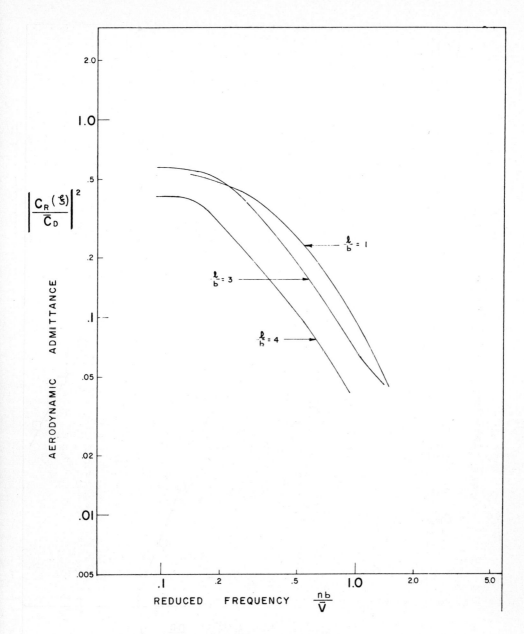

Fig.6. Aerodynamic admittance for rectangular discs determined from wind tunnel measurements.

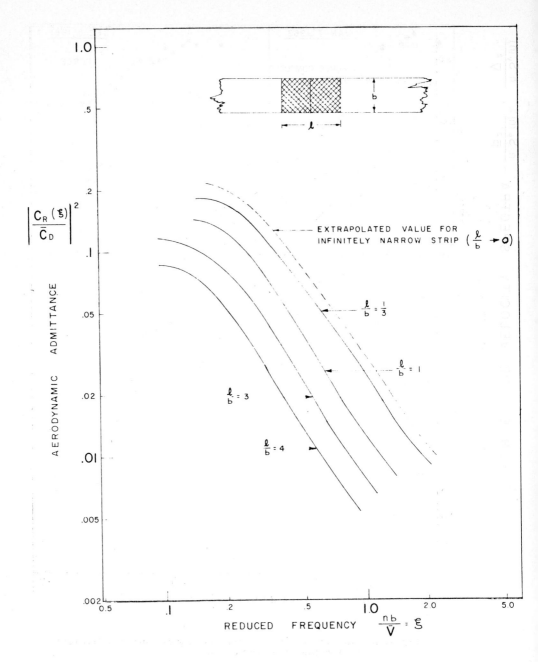

Fig.7. Aerodynamic admittance for panels of infinite flat plate

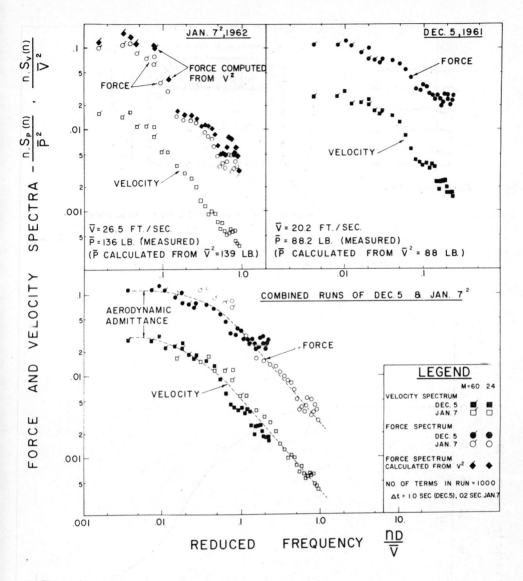

Fig. 8. Comparison of the wind velocity spectrum and the spectrum of the wind load on a 12 ft. square plate.

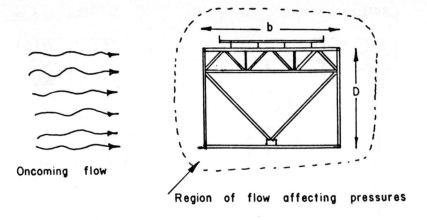

Fig.9. Diagram showing postulated region of flow affecting pressure on bridge truss.

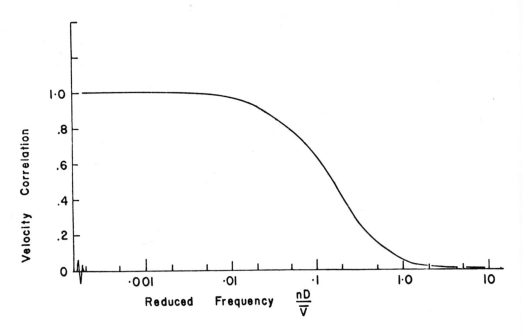

Fig.10. Velocity correlation over region of flow affecting pressure as function of reduced frequency.

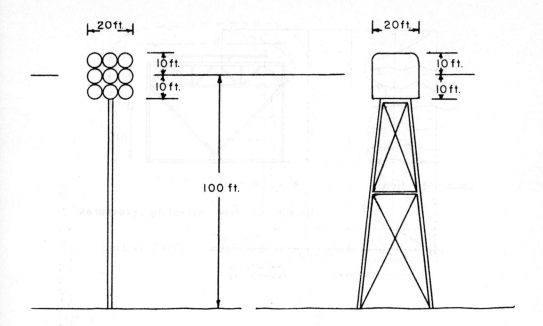

a. ARC LAMP

n_o = ·5 cy/sec

δ = ·10

b. WATER TOWER

n_o = 5 cy/sec

δ = ·10

Fig.11. Details of structures subjected to wind loads.

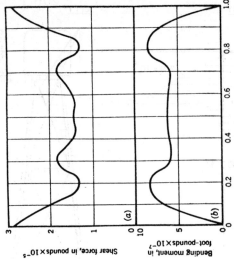
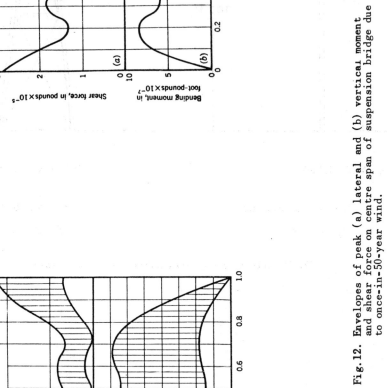

Fig.12. Envelopes of peak (a) lateral and (b) vertical moment and shear force on centre span of suspension bridge due to once-in-50-year wind.

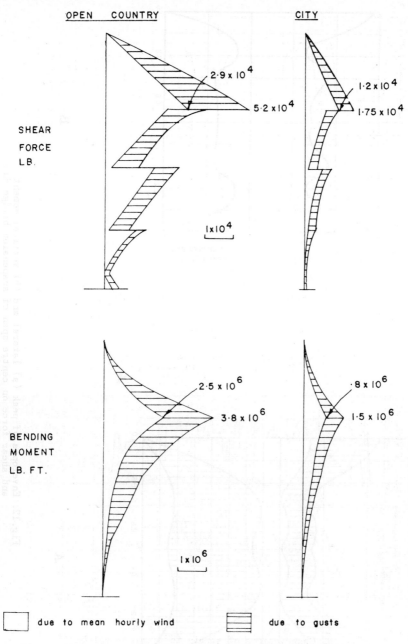

Fig. 13. Comparison of once-in-fifty year wind loading on a tall guyed mast under open country and city conditions (U_G = 80 mi/hr.: $\frac{1}{a}$ = 8 mi/hr.).

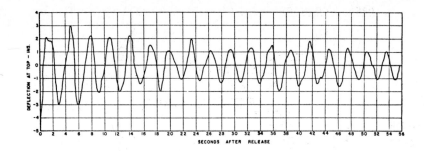

Fig. 14. Mechanical damping decay curve for 500 ft. guyed mast. (Obtained by photographic observation of mast-top after deflexion and sudden release). Damping rate found to be approximately 1% critical.

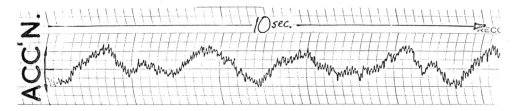

Fig. 15. Uncalibrated trace of acceleration at top of 1000 ft. guyed mast in a breeze.

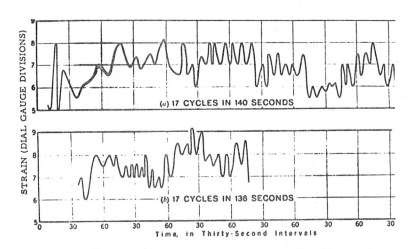

Fig. 16. Observations of strain in a corner column of the Empire State Building (25th floor) (after Rathbun).

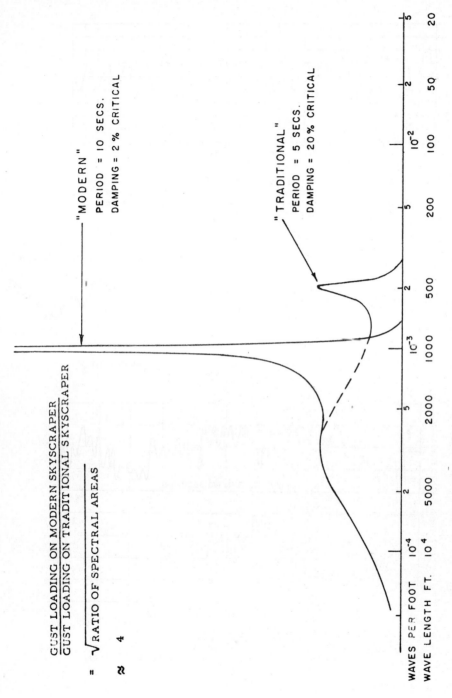

Fig. 17. Hypothetical response spectra for "traditional" and "modern" skyscrapers (fundamental mode of vibration only) V = 100 ft/sec. height 500 ft.

PAPER 18

THE RESPONSE OF STRUCTURES TO GUSTS

by

R. A. HARRIS
(Electrical Research Association, U.K.)

THE RESPONSE OF STRUCTURES TO GUSTS

by

R. I. HARRIS, B.A.
(Electrical Research Association, U.K.)

SUMMARY

THE limitations of existing methods of assessing wind loadings are briefly discussed, and the results of communication theory needed for the new statistical method of structural design are then introduced, together with a discussion of the meteorological results upon which the application of the new method depends. The pressure/velocity relationship which is used at present, is treated by an exact method, and the results show that approximations previously used are adequate in most cases. The need for an improved pressure/velocity relationship is briefly discussed, and some experiments at present being undertaken by the E.R.A. are outlined. The application of statistical methods to multi-degree of freedom systems is then introduced, and some experiments by the E.R.A. to find out the basic nature of wind structure are described. Finally, the need for adequate methods for the solution of non-linear problems is pointed out.

INTRODUCTION

In the course of its work on the potentialities of wind power, the Electrical Research Association developed new wind-measuring instruments (Rosenbrock[1]), and made extensive studies of wind structure, particularly at exposed sites (Wax[2], Tagg[3]). Some four years ago the Association's attention was directed specifically to the problem of wind loadings on structures, and a research programme was undertaken for a group of sponsors interested in various aspects of the problem. The work discussed in this paper represents some of the background and the results of this programme.

It is generally agreed that when any structure is designed, allowance must be made for wind loading, and in particular for the effect of gusts.

Until recent years the traditional methods for assessing wind loadings have been held to be satisfactory, possibly because designers have always been prepared to include large 'safety' or 'ignorance' factors in their calculations. The pressure of competitive conditions, steel shortages and military requirements have all, however, contributed to the production of lighter, more flexible structures, and to the need to reduce 'safety' factors to a minimum. The established methods of design would appear to have reached the limit of their development, since further progress is prevented by the unrealistic physical assumptions upon which they are based. The evolution of current methods and their limitations have been fully discussed by Davenport[4]. To summarise briefly, the methods at present in use are unsatisfactory for the following reasons:-

(i) The use of drag coefficients derived from wind tunnel tests, to relate the forces produced on a structure to the incident wind speed, is very questionable, when the wind speed is varying. There is a growing body of evidence (Davenport[5], Keulegan and Carpenter[6]) that there are extra contributions to the forces produced by the variation of the wind speed.

(ii) The power law for the variation of hourly mean wind speed with height must be adjusted to take account of changes in the roughness of various sites. Davenport[7] has shown that the value of the exponent in the power law can be satisfactorily related to the roughness of the terrain, and that the required information on the probability of obtaining a given hourly mean wind speed can be deduced from long term Meteorological data by the application of the methods of Extreme Value Statistics.

(iii) The failure to recognise that the natural wind is always 'gusty' or turbulent, in general, and therefore behaves randomly in space and time. This randomness implies that the problem of wind loading is a statistical one, and that any results which are derived can only be presented in statistical terms. It is indefensible to try to apply power laws to the variation of instantaneous wind speed with height.

(iv) It has long been felt that the 'gust factor' should bear some relation to the 'response time' of the structure under consideration, but no rational way of achieving this object has been evolved. The implication of this statement is that different sizes of structure will respond differently to the same wind environment. In other words, there is a dynamic interaction between the random, time varying, input forces acting on the structure, and the structure itself. It is therefore logically indefensible to consider wind loadings to be applied statically (Davenport[5], ERA[8], Harris[9,10]).

To overcome some of the above objections, we therefore require a method for treating the interaction of a random input with a dynamical system.

Suitable methods have been evolved by communications engineers for treating the behaviour of electrical noise in circuits (Wiener[11], Rice[12]), and have been adapted by Liepmann[13], Seiwell[14] and Panofsky[15] to problems in aeronautics, oceanography and meteorology, and by Rosenbrock[16] to the problem of gust loading on wind power plant, but it is only recently that their general use for the design of civil engineering structures has been advocated by Davenport[5] and Harris[9].

1. COMMUNICATION THEORY

The necessary results of communication theory will now be summarised, and their use for the solution of problems of wind loading on structures discussed. It should be emphasised that it is not possible in this paper to give more than a brief summary, and for further information the appropriate references should be consulted (Wiener[11], Rice[12], Blackman and Tukey[17], Crandall[18] and Lee[19]).

Consider a single degree of freedom dynamical system such as a mass/spring/dashpot system. This obeys a differential equation of the form:-

$$m\ddot{x} + c\dot{x} + kx = F(t) \qquad \ldots (1.1)$$

where m is the mass of the system
       c is the damping constant
       k is the spring constant
       x is the displacement of the system
and F(t) represents a time-varying force applied to the system.
It is usual to put (1.1) in standard form by letting

$$k/m = \omega_n^2, \quad c/m = 2\gamma\omega_n, \quad F(t)/k = f(t)$$

which gives:-

$$\ddot{x} + 2\gamma\omega_n \dot{x} + \omega_n^2 x = \omega_n^2 f(t) \qquad \ldots (1.2)$$

where ω_n is the natural frequency of vibration of the system in radians/unit time and γ is the dimensionless damping ratio.
Analysis of the dynamical behaviour of such a system usually proceeds along the following lines. The response of the system to a sinusoidal input is first considered by putting $f(t) = e^{j\omega t}$. From this the ratio of

the system output to the input is obtained; in this case the ratio is that of the displacement to the applied force. Since the complex exponential representation, $e^{j\omega t}$, for a sinusoidal input has been used, the ratio is also a complex quantity, $H(\omega)$ known as the complex frequency response of the system, where:-

$$H(\omega) = \frac{1}{1 - \left(\frac{\omega}{\omega_n}\right)^2 + 2j\gamma \frac{\omega}{\omega_n}} \qquad \ldots (1.3)$$

Note that $H(\omega)$ for a given ω depends only on the mechanical properties i.e., mass, damping, etc. of the system.

Two methods can be used to obtain the response of the system to transient inputs. In the first the response of the system, $h(t)$, to a unit impulse is first obtained, by considering the solution of (1.2) with $f(t) = \delta(t)$, the unit impulse function, and then the response to a general transient input $f(t)$ is found by the process of convolution i.e.

$$x(t) = \int_{-\infty}^{t} f(\tau) h(t - \tau) d\tau \qquad \ldots (1.4)$$

$$= \int_{0}^{\infty} h(\tau) f(t - \tau) d\tau \qquad \ldots (1.5)$$

(1.5) follows from (1.4) since for any physically realisable system $h(t) = 0$ if $t < 0$ as the output of any such system must follow, and cannot precede, the input.

The second method makes use of the properties of Fourier Transforms. Two functions $C(\tau)$ and $P(\omega)$ form a Fourier Transform Pair if they are related by:-

$$P(\omega) = \int_{-\infty}^{\infty} C(\tau) e^{-j\omega\tau} d\tau \qquad \ldots (1.6)$$

$$C(\tau) = \frac{1}{2\pi} \int_{-\infty}^{\infty} P(\omega) e^{j\omega\tau} d\omega \qquad \ldots (1.7)$$

The response of the system to a transient f(t) may be found by first finding the Fourier Transform of f(t). Let this be F(ω) where:-

$$F(\omega) = \int_{-\infty}^{\infty} f(t) e^{-j\omega t} dt \qquad \ldots (1.8)$$

by analogy with (1.6).

It can be shown that the Fourier Transform X(ω) of the system output x(t) is related to that of the input by:-

$$X(\omega) = H(\omega) \, F(\omega) \qquad \ldots (1.9)$$

and hence the output x(t) can be obtained from X(ω) by analogy with (1.7)

$$x(t) = \frac{1}{2\pi} \int_{-\infty}^{\infty} X(\omega) e^{j\omega t} d\omega \qquad \ldots (1.10)$$

It can also be shown that the impulse response h(t) and the complex frequency response H(ω) form a Fourier Transform Pair i.e.:-

$$H(\omega) = \int_{-\infty}^{\infty} h(\tau) e^{-j\omega \tau} d\tau \qquad \ldots (1.11)$$

$$h(\tau) = \frac{1}{2\pi} \int_{-\infty}^{\infty} H(\omega) e^{j\omega \tau} d\omega \qquad \ldots (1.12)$$

In order that this method shall be applicable, it is necessary for the input f(t) to satisfy certain mathematical conditions.

By either of the above methods the response of the system, both in amplitude and phase, to the given input is found.

Now consider the response of the system to a random input, or to an input with a random component. The input considered will usually have a non-zero mean, which is a constant, non-random or deterministic input. Since the system is linear, the effect of this component can be considered separately. When the response of the system to a random input f(t) is considered, it is found that no random input satisfies the necessary conditions for the application of the Fourier Transform method. If there is an experimental record of a random input, the response of the system in both amplitude and phase to this input can be computed, by application of (1.4) - the convolution method, but this proves to be a fruitless exercise. Since the input is random, any experimental record must be regarded as only a sample out of a large collection or statistical ensemble of such records that could equally well have been obtained, by recording the same variable. Under these conditions it is meaningless to obtain the amplitude and phase of the output from a given sample input. All that can be done is the calculation of some of the statistical properties of the output, given the statistical properties of the input, and the mechanical parameters of the system. If the mean value of the input is known the mean value of the output can be found by classical methods. Provided that the random input is both stationary and ergodic, other statistics of the input and output can also be related. If the input is stationary this implies that, although it is random, its statistical properties do not change with time, in other words samples of the input taken at various times conform to the same probability distribution. If the input is ergodic, this implies that statistical averages taken over a collection or ensemble of records are identical to those taken along one long record. In practice it is usually impossible to justify the assumption of stationarity and ergodicity except by the agreement of theory and experiment.

For a random input $f(t)$ the auto-correlation function $C_{11}(\tau)$ is defined by:-

$$C_{11}(\tau) = <f(t)f(t+\tau)> \qquad \ldots (1.13)$$

(In this expression, and in all that follows $<>$ denotes an average taken with respect to the time variable t) i.e. the value of $f(t)$ at some time t is multiplied by the value taken τ units of time later and this process is repeated for all values of t, and the resulting products averaged with respect to t. If the input is stationary and ergodic, the average so obtained is independent of t and may be computed either from one long record, or from a collection of pieces of record. Putting $\tau = 0$ in (1.13) gives:-

$$C_{11}(0) = <f^2(t)>$$

In other words, $C_{11}(0)$ is the mean square of $f(t)$. It can be shown that $C_{11}(\tau)$ satisfies the necessary conditions for the existence of a Fourier Transform $F_{11}(\omega)$

where

$$F_{11}(\omega) = \int_{-\infty}^{\infty} C_{11}(\tau) e^{-j\omega\tau} d\tau \qquad \ldots (1.14)$$

and hence

$$C_{11}(\tau) = \frac{1}{2\pi} \int_{-\infty}^{\infty} F_{11}(\omega) e^{j\omega\tau} d\omega \qquad \ldots (1.15)$$

Putting $\tau = 0$ in (1.15) gives:-

$$C_{11}(0) = \frac{1}{2\pi} \int_{-\infty}^{\infty} F_{11}(\omega) d\omega \qquad \ldots (1.16)$$

i.e. $F_{11}(\omega)$ represents the breakdown of the mean-square of the random process into frequency components. This problem was first considered for the case where $f(t)$ represented an electrical noise current. The average power dissipated by this current in a resistance of 1 ohm is $<f^2(t)>$, hence $F_{11}(\omega)$ is known as the power spectrum of the random process $f(t)$.

Now consider the application of the random input $f(t)$ to a dynamical system whose complex frequency response is $H(\omega)$. The output will be another random function $x(t)$, which will have an autocorrelation function $\chi_{11}(\tau)$ and a power spectrum $X_{11}(\omega)$

where:-

$$X_{11}(\omega) = \int_{-\infty}^{\infty} \chi_{11}(\tau) e^{-j\omega\tau} d\tau \qquad \ldots (1.17)$$

$$\chi_{11}(\tau) = \frac{1}{2\pi} \int_{-\infty}^{\infty} X_{11}(\omega) e^{j\omega\tau} d\omega \qquad \ldots (1.18)$$

and it can be shown that the input and output power spectra are related by:-

$$X_{11}(\omega) = |H(\omega)|^2 F_{11}(\omega) \qquad \ldots (1.19)$$

This is obviously analogous to the classical result (1.9). The difference is that the result involves the modulus of the complex response - in other words, only the relation of the amplitudes, and not the phases is considered. Hence, given a knowledge of either the autocorrelation function or the power spectrum of the input, plus a knowledge of the mechanical parameters of the dynamical system, the power spectrum of the output can be obtained. From this it is easy to obtain the mean square or variance of the output, since

$$< x^2(t) > = \chi_{11}(0) = \frac{1}{2\pi} \int_{-\infty}^{\infty} X_{11}(\omega) d\omega \qquad \ldots (1.20)$$

which is obtained from (1.18) by putting $\tau = 0$.

Thus there exists a method for obtaining the mean, mean square and autocorrelation function of the output. Knowledge of these quantities enables statistical predictions about the behaviour of the output to be made. If the output of the system is Gaussian i.e. the output has a probability distribution identical to the Normal Law of Errors, then the three quantities above completely determine its statistical properties. If the input is Gaussian, then since the system is linear, the output will also be Gaussian. In the more general cases where the input is not Gaussian, or the system is not linear, then the output is not Gaussian. If the system is lightly damped, however, then the output can often be shown to be asymptotically Gaussian, thus allowing the retention of useful results which can be obtained by assuming Gaussian behaviour.

2.   APPLICATION OF RESULTS TO WIND LOADING PROBLEMS

From previous work, it would appear that the wind velocity at a given height above ground level, and a given site, can be regarded as a Gaussian random variable with a non-zero mean. The variation of wind velocity contains components with a wide range of time scales from several days down to minutes or seconds, the former being associated with synoptic changes, and the latter being due to 'gustiness' or turbulence. Van der Hoven[20] has produced power spectra for the variation of wind velocity taken under a variety of atmospheric conditions, where the averaging time associated with the computation of the auto-correlation functions, and hence the power spectra, was long enough to include contributions from all the significant variations of wind velocity of all time scales. These spectra all show a common feature - there is little or no variation with a time scale of approximately one hour. This means that it is possible to take the averaging time for power spectra, etc., as one hour, when all variations of time scales less than one hour can be regarded as contributions to the mean square 'gustiness' or variance within the hour, while variations of time scales greater than an hour are regarded as slow drift in the value of the hourly mean wind velocity. In other words, variations of wind velocity can be regarded as quasi-stationary over a period of an hour, thus allowing the use of the results of Communication Theory. The use of an averaging period of one hour is very convenient, in that most long term meteorological data is available in the form of hourly means. Davenport[21] has demonstrated that it is possible to derive both the hourly mean and the mean square of the wind velocity at a given height above ground level, given a knowledge of the mean wind velocity at the standard reference height, or the gradient wind, together with a knowledge of the roughness parameter of the site. In the same paper, Davenport also showed that a universal power spectrum based on an hourly period, could, by use of the same parameters, be made to fit power spectra of 'gustiness' obtained experimentally from a large number of sites under various high wind conditions. The physical reason for obtaining an invariant shape of power spectrum is that under high wind conditions, the 'gustiness' is determined by the mechanical stirring effect of the rough boundary on the mean flow, and not by thermal processes.

Thus a structural designer, armed with a knowledge of the long term meteorological data for a site, together with a knowledge of the site roughness parameter which is obtainable by inspection, or by a few simple short term measurements on site, can work out by use of extreme value statistics the value of the hourly mean wind velocity which will occur with a given probability. From this hourly mean wind speed, the mean square 'gustiness' and power spectrum of gustiness for this 'highest hour' can also be worked out, and hence the designer can obtain the input power spectrum of wind velocity which the structure suffers. From a knowledge of the aerodynamics of the structure, the spectrum of wind velocity can be converted into an input force or pressure spectrum. By application of the Input/Output Theorem (Equation 1.19) the mean, mean square and autocorrelation of the output deflections can be calculated, and by assuming Gaussian behaviour an estimate can be obtained for the probability of a given deflection, and hence of the probability of failure of the structure. Associated with a given probability of failure is the notion of a return period which represents the average time required for the structure to fail, and it is obviously economically attractive that the return period should be related to the required life of the structure. The fact that failure can only be considered in terms of probabilities is a direct consequence of the random nature of the input wind velocity. Structural engineers have been used to designing for a 'maximum gust' which will never

be exceeded, so that structures capable of resisting such a gust will
never fail. Such notions are, however, fictitious since it is well known
in other fields of civil engineering, such as dam construction, that the
longer observations are continued, the greater is the value of the
'largest observed flood' which the structure will have to resist. The
reason that these ideas are not common in structural engineering is perhaps that, owing to 'safety factors' imposed for other reasons, structures
which have been erected have return periods for failure which are very
large, and correspondingly probabilities of failure which are very small.
Davenport[5] has shown how the above ideas may be utilised for the design of
two simple structures, and how the results can be presented in the form of
an equivalent gust factor. One important point is clearly demonstrated in
his paper - that the value of the equivalent gust factor is specific to a
given structure because of differences in dynamical behaviour, and hence
application of statistical methods for structural design as outlined above
makes it possible to adjust the gust factor for different structures in a
rational manner, and to take proper account of the random nature of the
winds.

3. THE RELATION OF PRESSURE SPECTRA TO VELOCITY SPECTRA

It has always been traditional to discuss wind loadings in terms of
incident wind velocities, while the quantity really required for wind
loading considerations, is the wind pressure. This is because meteorological data is always framed in terms of wind speeds. Critical examination,
however, shows that all meteorological instruments in common use respond
to wind pressure. In fact, the only types of anemometer which do not, are
the hot-wire and the acoustic anemometers. If the wind velocity is used
as a starting point then some form of functional relationship between
pressure and velocity must be assumed, in order to derive the required
pressures. The form of this relationship is by no means obvious -
traditionally, it has always been assumed that the same relationship between pressure and velocity obtains in gusty wind, as is measured by wind
tunnel tests. There is a growing body of evidence that extra effects due
to the varying nature of the incident flow must be considered. Davenport[5]
suggested tentatively the use of the following expression for the aerodynamic force on a body in varying flow:-

$$R_t = \frac{1}{2}\rho C_D V_t |V_t| + C_m \rho \frac{A_o}{D} \frac{dV_t}{dt} \qquad \ldots (3.1)$$

where R_t = force per unit area on the body at time t

V_t = velocity of the fluid at time t

ρ = fluid density

D = diameter of the body

A_o = reference area for the virtual mass $\left(\text{generally } \dfrac{\pi D^2}{4}\right)$

C_D = coefficient of drag

C_m = coefficient of virtual mass (including what is known as the additional or associated mass coefficient)

The first term on the right-hand side of equation (3.1) represents the form drag which is proportional in magnitude to the square of the wind velocity and directed in the same sense, while the second term represents the inertial reaction associated with acceleration of the fluid. Now suppose that V_t is a stationary, random function of time with mean value μ

i.e. $\qquad V_t = \mu + v(t) \qquad \ldots (3.2)$

where $v(t)$ represents the random time varying part of V_t, and has zero mean, and variance σ^2

i.e. $\quad <v(t)> = 0; \quad <v^2(t)> = \sigma^2$

and assume that the auto-correlation function of $v(t)$ is $C(\tau)$

i.e. $\quad <v(t)v(t+\tau)> = C(\tau) \qquad \ldots (3.3)$

It is convenient to use the normalised auto-correlation function defined by

$$\rho(\tau) = C(\tau)/\sigma^2 \qquad \ldots (3.4)$$

R_t will similarly be a stationary random function of time, and we shall denote the mean of R_t by $\bar{R}$ and the time-varying component by $z(t)$

i.e. $\qquad R_t = \bar{R} + z(t)$

and suppose that $z(t)$ has variance θ^2 and auto-correlation function $Z(\tau)$, that is:-

$$< z(t)z(t+\tau) > = Z(\tau) \qquad \ldots (3.5)$$

For convenience, put

$$\tfrac{1}{2}\rho C_D = A \qquad C_m \rho \frac{A_o}{D} = B \qquad \ldots (3.6)$$

Then

$$Z(\tau) = < \left(R_t - \bar{R}\right)\left(R_{t+\tau} - \bar{R}\right) > \qquad \ldots (3.7)$$

In his paper Davenport obtained an expression for $Z(\tau)$ by neglecting terms of the second order and higher, and derived in effect the following results:-

$$\bar{R} \doteqdot A\mu^2 \qquad \ldots (3.8)$$

$$Z(\tau) \doteqdot 4A^2\mu^2 C(\tau) + B^2 \frac{d^2 C(\tau)}{d\tau^2} \qquad \ldots (3.9)$$

In subsequent work Davenport also neglected the second term on the right-hand side of (3.9) which is derived from the auto-correlation of dV_t/dt. The derivation of this term involves no approximation and will be omitted from the discussion below.

The approximation involved in the derivation of (3.8) and (3.9) implies not only the neglect of higher order terms, but also the replacement of the term $AV_t|V_t|$ in the equation for R_t, by AV_t^2. If the wind behaves as a random variable, then for a small fraction of the time the instantaneous gust velocity $v(t)$ will be greater than the average flow velocity μ, and opposite in sign, so that instantaneously the wind velocity vector is directed against the mean flow direction. Existing standard meteorological instruments are not sensitive to wind direction i.e., they respond to the modulus of the wind velocity, or more strictly the wind-pressure vector, and until better measurements are available, it is difficult to say whether a significant physical effect exists. In the approximation used by Davenport, the instantaneous wind-pressure vector is assumed to have the same direction as the mean flow vector regardless of whether the direction of the instantaneous velocity vector is parallel or anti-parallel to the mean flow.

If (3.1) does represent the physical effects accurately, then it is possible to derive rigorous expressions for $\bar{R}$ and $Z(\tau)$ by using further results of communication theory - in this case, the theory of noise in an anti-symmetrical square law rectifier (Middleton[22]). A necessary assumption is that the 'gustiness' of the wind is a Gaussian random variable. It has already been stated that this is approximately correct, and hence errors due to this assumption are likely to be small.

Application of the appropriate theory gives the following results:

If $r = \sigma/\mu$

$$\bar{R} = A\mu^2 \left\{ r\sqrt{\frac{2}{\pi}} \exp\left(-\frac{1}{2r^2}\right) + (1 + r^2)\mathrm{erf}\left(\frac{1}{r\sqrt{2}}\right) \right\} \quad \ldots (3.10)$$

$$= A\mu^2 f_0(r) \quad \text{which serves to define } f_0$$

and

$$Z(\tau) = 4A^2\mu^2\sigma^2 \left\{ f_1(r)\rho(\tau) + f_2(r)\rho^2(\tau) \ldots + f_n(r)\rho^n(\tau) \right\} \quad \ldots (3.11)$$

where $f_1(r) \ldots f_n(r) \ldots$ are complicated functions involving exponentials and error functions of the ratio r.

In Davenport's treatment f_0 and f_1 are taken as unity, and all other f's are equated to zero. f_0, f_1, f_2 and f_3 are plotted in *fig.1*. The range of r of greatest practical significance is $0.2 < r < 0.5$. From the curves it will be seen that the approximation underestimates the mean pressure by between 4% and 25% in this range. The approximation for the f_1 term is remarkably accurate. The variance of the pressure is obtained by putting $\tau = 0$ in (3.11)

$$\theta^2 = 4A^2\mu^2\sigma^2 [f_1(r) + f_2(r) \ldots + f_n(r) + \ldots] \quad \ldots (3.12)$$

since $\rho(0) = \rho^2(0) = 1$ etc.

Hence for most practical cases only the contribution of the f_1 term to the variance of the pressure need be considered. To obtain the power spectrum (and hence the mean-square) of the output deflections, the power spectrum of the pressure is found by taking the Fourier Transform of $Z(\tau)$, and then multiplying by $|H(\omega)|^2$, the response function for the structure. It is inherent in the nature of Fourier Transforms that successive terms on the right-hand side of (3.11) make contributions to the power spectrum that extend over wider and wider bandwidths. Further, for a lightly damped system $|H(\omega)|^2$ is sharply peaked at the natural frequency of the structure, ω_n (*fig.2*) and hence most of the contribution to the variance of the output comes from the region of the input spectrum around ω_n. For most structures the natural frequency lies in the 'tail' of the gust spectrum, and it is in this region that the higher order terms predominate.

Consider a process for which

$$\rho(\tau) = \exp(-|\tau|) \qquad \ldots (3.13)$$

This gives a contribution to the power spectrum

$$F_1(\omega) = \frac{2}{1 + \omega^2} \qquad \ldots (3.14)$$

From (3.13) it follows that

$$\rho^2(\tau) = \exp(-2|\tau|) \qquad \ldots (3.15)$$

from which the contribution to the power spectrum is

$$F_2(\omega) = \frac{4}{4 + \omega^2} \qquad \ldots (3.16)$$

Hence the pressure spectrum in this case is

$$Z(\omega) = 4A^2\mu^2\sigma^2 \left[\frac{2f_1(r)}{1 + \omega^2} + \frac{4f_2(r)}{4 + \omega^2} + \text{higher terms} \right] \qquad \ldots (3.17)$$

Now suppose that in a particular case $\omega_n = 3$. This represents a typical example where ω_n lies in the 'tail' of the pressure spectrum. For $\omega = 3$ and, say, $r = 0.4$.

$$F_1(\omega) = 0.20 \qquad F_2(\omega) = 0.31$$

$$f_1(r) = 4.01 \qquad f_2(r) = 0.31$$

Thus the second order term contributes only 0.31/4.01 or rather less than 8% to the variance of the input pressure, but may contribute as much as $0.31^2/4.01 \times 0.2$ or 12% to the variance of the output deflection.

The issue is further complicated, as Davenport[5,23] has pointed out, by the necessity of multiplying the input pressure spectrum by an aerodynamic magnification factor. This takes account of dynamic variations in the drag coefficient C_D and the virtual mass coefficient C_m, which depend on the frequency of the variations in the incident flow. It also allows for the effect of loss of coherence of the velocity over an extended structure. Any structure must be insensitive to variations in velocity whose length scale is much smaller than the dimensions of the structure. The fact that the inclusion of this aerodynamic factor is necessary, is an indication that equation (3.1) does not completely represent the pressure/velocity relationship. In so far as (3.1) does represent the physical facts, the analysis above shows that for the majority of cases, the approximation used by Davenport is perfectly satisfactory, especially in view of the uncertainties in the aerodynamic relations. For cases where $r \sim 0.5$, or the structure is very lightly damped, then the exact theory will supply the necessary corrections. The method used to derive (3.10) and (3.11) can be applied to a variety of relations between pressure and velocity which can be expressed as polynomials in V_t, dV_t/dt and dV_t/dt^2, subject to the assumption of Gaussian behaviour, and certain restrictions on the differentiability of the random process. Hence the analysis should be useful with improved pressure/velocity relationships.

The derivation of a satisfactory way of relating anemometer readings to the forces experienced by structures is a vital link in the application of either the traditional or the new statistical design methods. For this reason some effort is being devoted to this problem in the current E.R.A. research programme. One experiment consists of a comparison between the output of a pressure-sensitive anemometer, and a velocity-sensitive hot-wire instrument placed in the same varying flow conditions. In a recent experiment the readings obtained from an array of pressure-sensitive anemometers are to be compared with the deflections produced in a simple pendulum made out of typical structural sections. These experiments represent an attempt to repeat in air the experiments done by Davenport[24] in a water tank, and should give information on both the dynamic variations of C_D and C_m, and the effects of incoherence.

4. EXTENSION TO MULTI-DEGREE OF FREEDOM SYSTEM

So far the problem has been discussed in terms of single degree of freedom dynamical systems. Most structures, and in particular, tall masts, are extended flexible bodies, and must, in general, be regarded as multi-degree of freedom systems. The treatment above can be extended to cover such systems by the usual methods of normal mode analysis. By these methods, the dynamic deflections of the structure are broken down into the superposition of a number of elementary deflections or normal mode shapes, each of which behaves, as far as its time-like properties are concerned, as a single degree of freedom system. Hence, the variance of the deflection, or of any quantity linearly related to the deflection, such as the slope or the curvature, can be found by a suitable summation over all the mode shapes, i.e., over all the equivalent single degree of freedom systems.

Although the problem can be discussed formally for the general case of a structure extended in three dimensions, as far as wind loading problems are concerned, it is only fruitful, given the present state of knowledge, to discuss the case where the structure is line-like, i.e., is extended in only one dimension. Fortunately this covers a large number of structures of interest. The assumption of a line-like structure means that all the properties of the structure such as mass, stiffness, etc., and the deflection can be regarded as functions of only one space variable. For a line-like structure it is also meaningful to define a drag coefficient per unit length, so that the vertical wind velocity profile can be related to the vertical profile of the force per unit length produced in the structure. Where the structure is extended in two dimensions as in the case of a large building, no such relation can be deduced at present. A linear elastic structure will obey a partial differential equation of the form:-

$$m(z)\frac{\partial^2 x}{\partial t^2} + c(z)\frac{\partial x}{\partial t} - y(z)D_z\{x\} = p(t; z) \qquad \ldots (4.1)$$

where z is the independent space variable

t is the time variable

x(t; z) is the displacement of the system

m(z) is the mass/unit length of the system

c(z) is the damping/unit length of the system

$D_z\{\ \}$ is a linear homogeneous differential operator which characterises the stiffness of the system

e.g. for a beam $D_z\{\ \} \equiv \dfrac{\partial^4}{\partial z^4}$

y(z) characterises the variation in stiffness/unit length of the system i.e., due to changes in section or in elastic modules.

p(t; z) represents the input force/unit length which is a random function of space and time.

Since $y(z) \neq 0$, division by $y(z)$ gives:-

$$K(z)\frac{\partial^2 x}{\partial t^2} + b(z)\frac{\partial x}{\partial t} - D_z\{x\} = \frac{p(t; z)}{y(z)} \qquad \ldots (4.2)$$

where $\quad K(z) = m(z)/y(z)$

$\qquad b(z) = c(z)/y(z)$

Note that p(t; z) is considered to be random with zero mean. In general, in wind loading problems the force per unit length will have a non-zero mean, but the contribution of this to the output deflection of the structure can be treated separately by classical methods, as for the single degree of freedom case. For p(t; z) a general cross-correlation function $P_{12}(\tau; z, z')$ is defined by:-

$$P_{12}(\tau; z, z') = \langle p(t; z)p(t + \tau; z') \rangle \qquad \ldots (4.3)$$

i.e. the force per unit length at the point z in the structure and time t, is multiplied by the force per unit length at the point z' taken τ units of time later. The resulting product is then averaged for all values of t. $P_{12}(\tau; z, z')$ satisfies the necessary conditions for the existence of a Fourier Transform $P_{12}(\omega; z, z')$ where:-

$$P_{12}(\omega; z, z') = \int_{-\infty}^{\infty} P_{12}(\tau; z, z') e^{-j\omega\tau} d\tau \qquad \ldots (4.4)$$

with the inverse relation:-

$$P_{12}(\tau; z, z') = \frac{1}{2\pi} \int_{-\infty}^{\infty} P_{12}(\omega; z, z') e^{j\omega\tau} d\omega \qquad \ldots (4.5)$$

Now if the output deflection of the structure at the point z and time t is x(t; z), this will also be a random function of both space and time. The auto-correlation of the deflection at the point z will be:-

$$\mathcal{X}_{11}(\tau;\ z) = \langle x(t;\ z)x(t+\tau;\ z) \rangle \qquad \ldots (4.6)$$

which will have a corresponding power spectrum $X_{11}(\omega;\ z)$ given by:-

$$X_{11}(\omega;\ z) = \int_{-\infty}^{\infty} \mathcal{X}_{11}(\tau;\ z) e^{-j\omega\tau} d\tau \qquad \ldots (4.7)$$

with the inverse relation

$$\mathcal{X}_{11}(\tau;\ z) = \frac{1}{2\pi} \int_{-\infty}^{\infty} X_{11}(\omega;\ z) e^{j\omega\tau} d\omega \qquad \ldots (4.8)$$

By analogy with (1.16), the variance or mean-square of the deflection at the point z is:-

$$\langle x^2(t;\ z) \rangle = \mathcal{X}_{11}(0;\ z) + \frac{1}{2\pi} \int_{-\infty}^{\infty} X_{11}(\omega;\ z) d\omega \qquad \ldots (4.9)$$

When normal mode analysis is applied to the system, the following result is obtained:-

$$X_{11}(\omega;\ z) = \sum_n \sum_m \frac{A_n(z)A_m(z)}{K_n\omega_n^2 K_m\omega_m^2} H_n^*(\omega) H_m(\omega) C_{nm}(\omega) \qquad \ldots (4.10)$$

where

$$C_{nm}(\omega) = \int_0^L \int_0^L \frac{A_n(z)A_m(z')P_{12}(\omega;\ z,z')}{y(z)\ y(z')} dz\ dz' \qquad \ldots (4.11)$$

$A_n(z)$, $A_m(z)$ are the nth and mth mode shapes.

$H_n(\omega)$ is the complex frequency response of the nth mode

$H_n^*(\omega)$ is the complex conjugate of $H_n(\omega)$

ω_n is the natural frequency of the nth mode

K_n is the normalization constant of the nth mode

i.e.
$$K_n = \int_0^L K(z) A_n^2(z) dz$$

The structure is assumed to stretch from $z = 0$ to $z = L$.

Hence knowing the input pressure cross-spectrum and the mechanical properties of the structure, the double summation as indicated in (4.10) is performed, followed by integration with respect to frequency, to obtain the variance and auto-correlation of the deflection, at any point in the structure. From these quantities statistical predictions can be made of the probability that a given deflection will occur. The result above was given in slightly different form by Powell[25]. For most mechanical systems, the damping is small, and in this case only the 'diagonal' terms in (4.10) need be considered, i.e.:-

$$X_{11}(\omega; z) \doteq \sum_n \frac{A_n^2(z)}{K_n \omega_n^4} \left| H_n(\omega) \right|^2 C_{nn}(\omega) \qquad \ldots (4.12)$$

where

$$C_{nn}(\omega) = \int_0^L \int_0^L \frac{A_n(z) A_n(z') P_{12}(\omega; z, z')}{y(z) y(z')} dz \, dz' \qquad \ldots (4.13)$$

This development is due to Thomson and Barton[26].

Examination of the nature of the terms in (4.12) shows that the $C_{nn}(\omega)$ term represents the interaction of the spatial properties of the structure with the spatial properties of the input pressure, while the $\left| H_n(\omega) \right|^2$ term represents the further interaction of the time-like properties of the system and the input. All the necessary properties of the wind are contained in $C_{nn}(\omega)$, and the solution of the problem of wind loading is reduced to the measurement and presentation of $C_{nn}(\omega)$ in a form useful for design purposes. Unfortunately $C_{nn}(\omega)$ depends upon $A_n(z)$, $y(z)$ and the distribution of area and drag for the structure, and hence any given $C_{nn}(\omega)$ is specific to the particular structure whose modes and stiffness functions are used to evaluate the integrals. In certain cases where the wind structure has special properties, the general expressions (4.12) and (4.13) can be simplified. Davenport[27] has recently treated the case of a suspension bridge, where because the structure is horizontal, $P_{12}(\omega; z,z')$ can be regarded as a function only of ω and $\xi = z-z'$, which considerably reduces subsequent computation. For vertical structures, however, it is uncertain whether similar simplifications can always be made.

An alternative approach is being tried in the current E.R.A. programme in order to achieve generality. Suppose that the force per unit length in the structure is related to the velocity at the same point, by the conventional square law

$$\text{i.e.} \quad p(t; z) = \frac{1}{2}\rho C_D(z) A(z) V^2(t; z)$$

where $C_D(z)$ is the drag coefficient/unit length at the point z

$A(z)$ is the area/unit length at the point z

ρ is the air density

$V(t; z)$ is the wind velocity at the point z and time t.

As remarked earlier, this relation requires some modifications in practice, but these modifications do not invalidate the present treatment. Then

$$P_{12}(\omega; z,z') = \frac{1}{4}\rho^2 C_D(z) C_D(z') A(z) A(z') W_{12}(\omega; z,z') \quad \ldots (4.14)$$

where $W_{12}(\omega; z,z')$ is the cross spectrum of the square of the velocity.

Hence

$$C_{nn}(\omega) = \frac{\rho^2}{4} \int_0^L \int_0^L \frac{A_n(z) C_D(z) A(z)}{y(z)} W_{12}(\omega; z,z') \frac{A_m(z') C_D(z') A(z')}{y(z')} dz\, dz'$$

$$\ldots (4.15)$$

A generalized $C_{nn}(\omega)$ is computed by replacing the functions $A_n(z) C_D(z) A(z)/y(z)$ in (4.15) by a set of orthogonal polynomials - in this case, shifted Chebyshev Polynomials of the first kind $T_n^*(z)$. For a given structure the function $A_n(z) C_D(z) A(z)/y(z)$ is also obtained as an expansion in the same polynomials and the $C_{nn}(\omega)$ specific to the structure can then be derived, by multiplying matrices consisting of the coefficients of the appropriate polynomials. To fit in with this approach, it is advantageous to arrange experiments so that measurements are taken at the zeros of one of the polynomials.

On one of the masts at the G.P.O. Rugby Radio Station an array of instruments has been installed covering a vertical range of 0 - 300 ft. *(fig.3)*. The instruments used are pressure-sensitive anemometers with a true cosinusoidal polar diagram, and hence the instruments measure the component of wind pressure in a given direction. The readings from these instruments are sampled simultaneously, and recorded in digital form on magnetic tape, for subsequent processing by an electronic computer. The results from this experiment will provide data on a number of aspects of the wind loading problem. The invariance of the one point power spectrum with height can be checked more adequately than has been possible with the measurements previously available. The invariance under various conditions of atmospheric stability can also be checked. As remarked previously these ideas are vital to the application of the statistical design method. The pressure cross-spectrum $P_{12}(\omega; z,z')$ can be examined to see if approximations of the type used for the horizontal case, are also applicable to the vertical case, and hence the necessity and utility of the Chebyshev approach can be tested. At the time of writing, insufficient results are available to answer any of these points, and in particular the Chebyshev method must be regarded as conjectural. It is intended to extend the measurements beyond 300 ft as soon as sufficient experience has been gained.

As there is also considerable interest among manufacturers and users of large radar installations in the horizontal distribution of the wind, an experiment is also in progress to measure the horizontal correlation of wind pressure across a wide front. At the E.R.A's Cranfield Field Station, a line of six masts has been set up at right angles to the prevailing wind, with anemometers at 10 metres above ground level *(fig.4)*. The masts are spaced non-uniformly at intervals of 1, 3, 6, 5 and 2 units of 45 ft, so that the total length of the line is 765 ft. The reason for using this spacing is that it allows estimates of the correlation to be obtained for all separations from 1 to 17 units, with the exception of 8 and 12 units, by selecting appropriate pairs of instruments, and thus represents an economical layout for obtaining a detailed estimate of the horizontal correlation. The instruments used on these masts are identical to those used for the Rugby experiment. The results will similarly be recorded on magnetic tape and processed by computer.

5. NON-LINEAR SYSTEMS

The analysis in this paper, and in all previous work published on the statistical method of assessing gust loadings, is applicable only to linear dynamical systems i.e., to structures whose equations of motion are linear differential or partial differential equations. As far as this type of structure is concerned, the present state of knowledge allows us to see in outline the methods for assessing wind loadings, although many points still need proper experimental confirmation. Unfortunately many structures of principal interest are non-linear in behaviour, and here the position is less clear. The stayed mast is non-linear because of the action of the stay catenaries and the effects of buckling, and some recent occurrences of oscillations on a stayed mast in North Wales[28], suggest that the various linearisation methods commonly used, are inadequate to describe the behaviour of the structure. Closely related problems concern oscillations induced by various forms of instability. These are normally regarded as steady-flow problems, but the role of gusts in starting, modifying and maintaining oscillations has recently been considered by Davis et al[29]. A considerable volume of work has been done by communications engineers on the random excitation of non-linear single degree of freedom systems (Crandall[30]), and some of the results would appear to be of potential value for problems related to structures. In view of the particular importance of stayed masts, the E.R.A. is currently taking part in an experiment to study the non-linearities of catenaries. A 200 ft span of ¼" O.D. Aluminium wire has been set up with facilities for the measurement of tension and displacement at various points in the span. Various boundary conditions can be applied - the catenary can be forced at a wide range of amplitudes and frequencies by a motor and flywheel assembly, or be provided with fixed ends, or with one fixed end and one end controlled by a linear spring. The object of this experiment is the development of suitable methods for describing the non-linearities. Since very little work is available on the random motion of systems governed by non-linear partial differential equations, the most fruitful form of attack on non-linear wind loading problems seems, at the moment, to be a simulation technique. The input to the system will be known from wind structure measurements, and by simulating the required input power spectra and dynamical system, with the aid of either an analogue or a digital computer, a large number of sample outputs can be examined, and hence the required statistics can be deduced. This is obviously a field for further study.

ACKNOWLEDGEMENT

The author would like to thank his colleagues in the Electrical Research Association, the Research Committee of Sponsors for Research Programme SP/1, and Mr. T. J. Poskitt, for the help and encouragement that he has received. His thanks are also due to the Director of the E.R.A. for permission to publish the paper.

REFERENCES

The confidential reports listed are those presented to the Research Committee for the E.R.A. Sponsored Programme SP/1 - "Wind Loading on Structures".

1. ROSENBROCK, H. H. The design and development of three new types of gust anemometer. *E.R.A. Technical Report C/T106*, 1951.
2. WAX, M. P. An experimental study of wind structure. *E.R.A. Technical Report C/T114*, 1956.
3. TAGG, J. R. Wind data related to the generation of electricity by wind-power. *E.R.A. Technical Report C/T115*, 1957.
4. DAVENPORT, A. G. Wind loads on structures. *Tech. Rept. 88, Division of Building Research, Nat. Res. Council, Canada*, March, 1960.
5. DAVENPORT, A. G. The application of statistical concepts to the wind loading of structures. *Proc. Instn. civ. Engrs.*, 1961, $\underline{19}$, 449.
6. KEULEGAN, G. H. and CARPENTER, Ll. H. Forces on cylinders and plates in oscillating fluid. *J. Res., nat. Bur. Stds.*, 1958, $\underline{60}$, 5, 423.
7. DAVENPORT, A. G. Rationale for determining design wind velocities. *Proc. Amer. civ. Engrs., J. struct. Div.*, 1960, $\underline{86}$.
8. GOLDING, E. W., GIMPEL, G., MORRISON, J. G. and TAGG, J. R. The influence of wind on the design of tall structures. *E.R.A. Confidential Report SP1/T1a*, March, 1959.
9. HARRIS, R. I. E.R.A. Confidential Report SP1/T4, September, 1960.
10. HARRIS, R. I. Wind pressures on large structures. *'Electrical Review'*. 20th July, 1962.
11. WIENER, N. *Extrapolation, interpolation and smoothing of stationary time series.* Technology Press, Cambridge, Mass. and Wiley, New York, 1949.

12. RICE, S. O. Mathematical analysis of random noise. *Bell Sys. Tech. Journ.* 23, 282, 1944 and 24, 46, 1945. Reprinted in *"Selected Papers on Noise and Stochastic Processes"*. ed. Wax, Dover Publications, New York, 1954.
13. LIEPMANN, H. W. On the application of statistical concepts to the buffeting problem. *J. aero. Sci.*, 1952, 19, 12, 793.
14. SEIWELL, H. R. The principles of time series analysis applied to ocean wave data. *Proc. nat. Acad. Sci.*, 1949, 35, 518.
15. PANOFSKY, H. A. Meteorological applications of power spectrum analysis. *Bull. Amer. Meteorological Soc.*, 1955, 36, 163.
16. ROSENBROCK, H. H. Vibration and stability problems in large wind turbines having hinged blades. *E.R.A. Technical Report C/T113*.
17. BLACKMAN, R. B. and TUKEY, J. W. *The measurement of power spectra from the point of view of communications engineering*. Dover Publications, New York, 1959.
18. CRANDALL, S. H.(ed). *Random vibration*. Technology Press, Cam., Mass. and Wiley, New York, 1958.
19. LEE, Y. W. *Statistical communication theory*. Wiley, New York, 1961.
20. VAN DER HOVEN, I. Power spectrum of horizontal wind speed in the frequency range from 0.0007 to 900 cycles per hour. *J. Meteor.* 1957, 14, 160.
21. DAVENPORT, A. G. The spectrum of horizontal gustiness near the ground in high winds. *Quart. J. roy. meteor. Soc.*, 1961, 87, 372, 194.
22. MIDDLETON, D. *An introduction to statistical communication theory*. McGraw Hill, New York, 1960.
23. DAVENPORT, A. G. Discussion on paper published in Proceedings in August, 1961. *Proc. Instn. civ. Engrs.*, 23, pp.143-146, September, 1962.
24. DAVENPORT, A. G. A statistical approach to the wind loading of the tall mast and suspension bridge. *Ph.D. thesis*, University of Bristol, Dept. of Civil Engineering, 1961.
25. POWELL, A. Response of structures to jet noise. *"Random Vibration"* - see (18) above.
26. THOMSON, W. T. and BARTON, M. V. The response of mechanical systems to random excitation. *J. appl. Mechs.*, 1957, 24, 2, 248.
27. DAVENPORT, A. G. The response of slender, line-like structures to a gusty wind. *Proc. Instn. civ. Engrs.*, 1962, 23, 389.
28. B.I.C.C. *Private communication*.
29. DAVIS, D. A., RICHARDS, D. J. W. and SCRIVEN, R. A. Investigation of conductor oscillation on the 275 kV crossing over the Rivers Severn and Wye. *Proc. Instn. elec. Engrs.*, 1963, Paper No. 4102 P, 110, No.1, 205.
30. CRANDALL, S. H. Random vibration of systems with non-linear restoring forces. *M.I.T. Report AFOSR 708* under Contract 49(638) - 564, June, 1961.

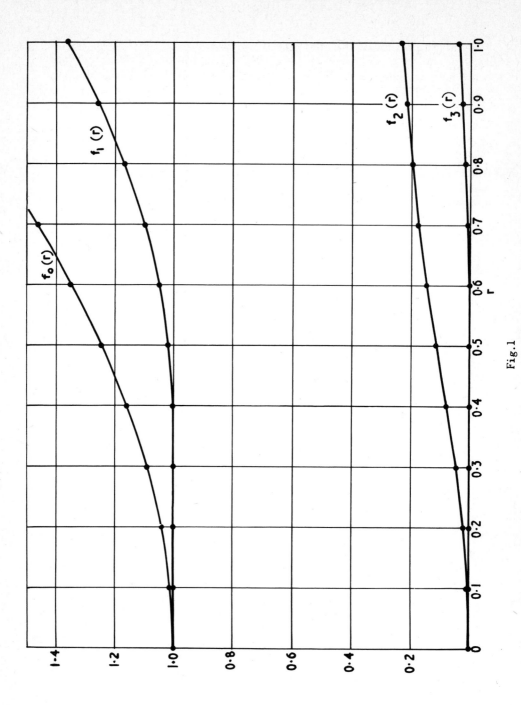

Fig.1

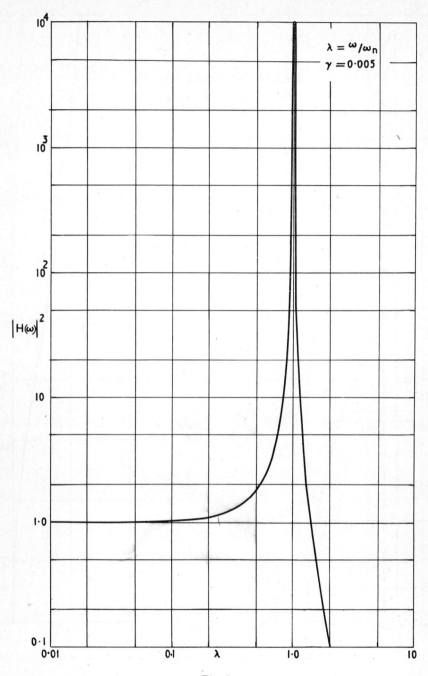

Fig.2

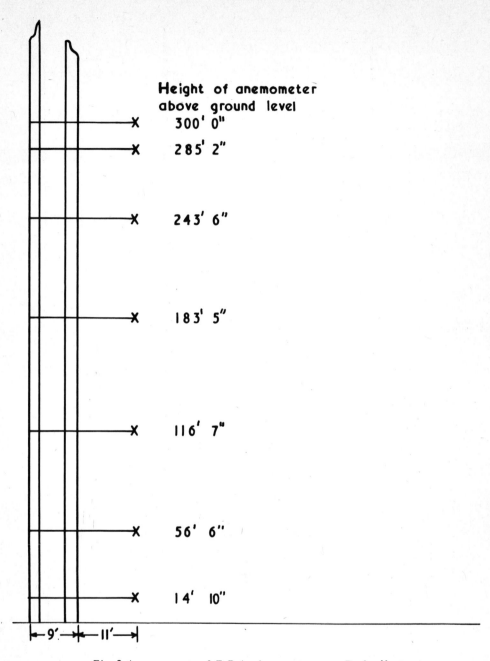

Fig.3 Arrangement of E.R.A. Anemometers on Rugby Mast

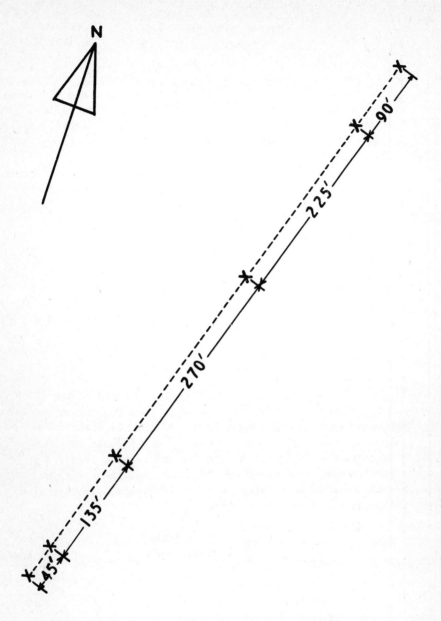

Fig.4 Layout of E.R.A. Horizontal Correlation Experiment

DISCUSSION ON PAPERS 9 and 18

MR. FALKINER-NUTTALL. In the United Kingdom the electricity regulations for the last 30 years have made no allowance for the increase of wind loading due to ice-coating on the structure and there have been only one or two recent failures. Radio and television structures are designed with such an allowance and the Admiralty allow a safety factor of 6. He thought that there should be some co-ordination between the different bodies to standardise the requirements.

MR. FISHER protested that the information which is being provided to write codes of practice comes from meteorologists and aerodynamicists and he felt that these specialists should be aware of what the ordinary structural engineer wants. He drew a distinction between the structure of a building and the fabric, the former being the framework. Nearly all reported failures had been failures of the fabric and not of the structure. To assess the effect of wind on a structure the first requirement is to put strain or force gauges on the structural members. Pressure gauges in the glass of the building give no direct assessment of the wind effect on the structure. It was essential to have a simple code and one in which the structure and the wind were not considered in isolation and in which differentiation is made between different types of buildings.

PROFESSOR MACKEY. In Hong Kong, storm records over the past 72 years show mean hourly wind speeds ranging from 28 to 90 m.p.h. Over the same period the extreme gust factor ranged from 1.43 to 2.21 with a mean value of 1.73. Typhoons (that is, storms of over 75 m.p.h.) are supposed to occur in Hong Kong once in 15 years. An extreme typhoon occurred in 1937 with a maximum recorded wind speed of 167 m.p.h. and these extreme typhoons are supposed to occur about once in 100 years. In fact since 1957 the following typhoons have occurred:

Date	Name	Mean hourly wind speed (m.p.h.)	Maximum gust speed (m.p.h.)
Sept. 1957	GLORIA	68	116
June 1960	MARY	58	119
May 1961	ALICE	50	103
Sept. 1962	WANDA	78	167

For typhoon MARY the maximum mean wind speed for various "gust" durations were

Duration	Mean Speed (m.p.h.)
1 hour	58
5 minutes	61
1 minute	70
30 seconds	80
10 seconds	92
5 seconds	95

Most of our information dealt with the action of dry wind, but the action of wind with rain might be significantly different. In Hong Kong water virtually beats horizontally against the structures, and in certain conditions it runs up the curtain walling rather than down! It may be of interest to note that during typhoon GLORIA the only major scaffolding collapse was with tubular steel rather than with bamboo scaffolding.

PROFESSOR DAVENPORT (in reply). To study the problem of wind loading in terms of discrete gusts is not going to provide the final solution because time history of gusts is also of consequence. The reason for the better behaviour of the bamboo scaffolding as instanced by Professor Mackey, might be that the bamboo scaffolding had lower frequiencies and greater damping than had the steel scaffolding.

MR. HARRIS (in reply) agreed with Mr. Fisher that the wind and the structure should be considered together. What Professor Davenport has shown is that the gust factors that are obtained by applying his theory depend on the dynamic nature of the structure, and thus, in a sense, the wind loading derived is specific to the structure under consideration. This fact makes it difficult to reduce results to equivalent static wind pressures. It might be said that the difference between engineers of the present and of the 19th century is not one of ability but that we now have computers. Computers will enable the more complex computations required for the calculation of dynamic response of structures to gusting winds. Design of masts, for instance, can then take place by an iterative process in which the answers are examined and then the structure is stiffened up in various places, until a satisfactory design is reached. It may take a long time before we arrive at this stage of development.

WRITTEN DISCUSSION
PAPERS 19, 14, 9 and 18

DR. WYATT. When engaged in the design of a number of structures where gust loading was of paramount importance, he had become acutely aware of the unsatisfactory logical framework hitherto available for the correlation and codification of such information as was available. As it will presumably be some time before the results of full application of the methods expounded by Professor Davenport can come into general use, he had looked for immediate benefits by improving the framework for simple design rules.

In particular there is the determination of the effective gust duration: this is unsatisfactory in the existing British Standard CP3 Chapter V. It is to be hoped it will now be possible to relate this logically to the size and natural frequency of the structure. The best guidance hitherto has been Professor Sherlock's work, but his conclusion that a gust duration equivalent to some ten times the size of the structure is necessary to allow the force to reach the quasi-steady value would appear to require modification in the light of the recent experiments described in the papers. On this subject the writer wished to ask M. Esquillan what gust duration was considered in determing the French rules NV63 (*fig.1* of paper 19).

Professor Davenport has already shown the importance of damping: this should be reflected in a reduction factor for loading on massive buildings of "old fashioned" construction where a considerable degree of structural damping is assured. Simple rules are likely to be less satisfactory, however, in dealing with structural members carrying loads primarily dependent on the difference between wind force on two parts of the structure, which occur in guyed masts, "Eiffelised" towers, radiotelescopes, etc. The Danish specification as given by Dr. Jensen and Mr. Frank in paper 14 that 1/3 of the total be considered as a live load, irrespective of size or any other factor, is a severe simplification, although any guidance is better than none. In Britain the basis of design of electricity transmission towers is laid down by law but without any reference to such a factor!

MR. FOSTER. The range of problems concerned with wind loadings and wind spectrum being discussed at this Conference is of interest to the firm I am associated with in connection with a new and novel type of large structure. I refer to the dry cooling tower used in conjunction with thermal power stations where no large make-up quantities of cooling water are required, and the first unit in this country, of 120 MW, has now been operating for some time at the C.E.G.B's Rugeley Generating Station.

The tower itself is 350' high and has a base dia. of 325'. The cooler elements 48' high are situated round the base, and the tower itself operates by natural draught. If the new system proves to be both a practical and economic proposition, much larger towers are possible in the future, and one gets concerned therefore about wind effects both on the performance and the stress loading in the tower shell and the cooler steelwork.

The tower itself can be regarded rather crudely for aerodynamic purposes as a large vertical open-ended cylinder of aspect ratio about unity and with about 15% of its base perforated. The flow pattern both around and inside the tower in a wind is therefore rather complex, and one difficulty is that the Reynolds numbers involved are far higher than have ever been reached in a wind-tunnel even on an infinite cylinder. We are talking about 10^8 for Rugeley in a 50 m.p.h. wind, and values of twice this are likely on larger projects.

The structural design of both the tower shell and the coolers themselves is very much dependent upon the specific wind loading assumed. We have carried out tests on a model of the Rugeley tower under a steady uniform wind to obtain our $\pm C_p$ variations, but several papers at this Conference have emphasised that far more factors are involved than were previously accepted, and that blanket wind pressures deduced from a single maximum possible steady wind are hardly good enough.

The papers have clarified several matters concerning the influence of terrain, maximum wind levels, velocity gradients, return periods, gust ratios, etc., which are very helpful. I would like to ask however, in connection with these large towers, whether the new spectral approach reviewed in Papers 9 and 18 could be used to obtain some idea of the fluctuating loads due to gusting? Is the atmospheric horizontal gusting of a large enough scale and intensity to envelope the tower and give it severe blows in spite of its inherent structural damping, and are these likely to be more or less severe than those resulting from fluctuations due to vortex shedding at the rear of the tower? One must assume a natural period of 1 or 2 secs for the structure, whilst the major shedding frequency (assuming a Strouhal No. of 0.3), gives durations of 10 to 30 secs.

MR. GIDWANI. Most of the Contributors of papers at the Conference were obviously engaged principally in research. Whilst their reports were most useful, I would have welcomed more contributions from practicing Engineers actually engaged in design and construction of structures affected by wind, and the benefit of their experience. In my view insufficient attention was given to practical aspects of wind effects on tall buildings. Considerable research work seems to have been conducted and data collected on smaller buildings and other structures, but there is little easily accessible and applicable data available for tall buildings say more than 200 feet high, and there does not appear to be full appreciation of Structural Engineers' problems. Unless the Research worker appreciates these problems and the results of research are easily available to the Engineer in a concise and easily applicable form, he is likely to design by intuition or judgment or intelligent guesswork. Most Engineers forget their higher mathematics by the time they reach a position of making basic decisions for important structures.

Referring to statistical and probability approach, surely this is not applicable to Bridges and Civil Engineering structures although one may take calculated risks for aeroplanes and possibly pylons, overhead power transmission lines and some sea defence works, etc. The Engineer for a building, bridge or other public structures has a moral and legal responsibility to his Clients, users or tenants and the general public, particularly where human life may be concerned. You can build a prototype of an aeroplane or a pylon and test to destruction but not for an important bridge or a tall building. I would certainly not like to tell a Client that his building may collapse in a storm which may occur only once every 100 years, but that fatal year may be any year - possibly next year. I would not want to live in my house if I thought I had a one in a 100 chance. This explanation is no comfort to the sufferers of Sheffield storm damage in 1962, so poignantly illustrated by Professor Page.

Of course it is not possible to design against all unknown eventualities or accidents that may occur, but one should design buildings, bridges, and similar structures for any forces of nature that can be reasonably foreseen. Dr. Blair's reference to a plane crashing on a house is irrelevant; you do not design windows of a house against a boy throwing stones at it.

Mr. Fisher rightly pointed out the problems of the design Engineer but, in my view, both Dr. Blair and Mr. Fisher appear too complacent about the effects of wind and suspicious of research. The destructive effects of brute force of wind on topography and man's creations are much too apparent. Many structures have suffered from storm damage in spite of the design being in accordance with Codes of Practice or other accepted Standards. If there have not been more failures of buildings due to wind forces, it is not due to our full understanding of wind forces but due to inherent stability of heavy buildings and perhaps good luck. But in recent years the increase in the size of structures and use of modern materials and methods of construction have made the wind effects more critical and the need for research more urgent. We cannot wait to learn by experience or failures alone. In the case of tall buildings, a proper study of wind effects and appropriate allowance in design would affect only the structural arrangement and details and should not increase the cost appreciably.

It was obvious from many contributions that the effect of gusts, turbulence, etc. were more important than just static wind pressure, but I have not yet seen a practical method of making allowance for these in design. Obviously these will depend on the surrounding topography, shape of building and other characteristics peculiar to each individual case. Can they be generalized? There appears to be no means of determining these effects except by a wind tunnel test on an exact model which may not be practicable or possible in every case. Nor is it yet certain that wind tunnel tests at present are representative of actual conditions and that results of such tests could be relied upon. It is not yet possible to produce a model of a traditional building with exact characteristics, including damping factor, gravity scale, stiffness, etc. to allow for effects of cladding, partitions, etc. or to simulate surrounding conditions and wind characteristics of the actual site in the wind tunnel. It may well be that because of these reasons wind tunnel tests give misleading results for buildings, although they have obviously proved their reasonable accuracy and usefulness for suspension bridges, pylons, stacks or similar.

Professor Davenport suggested site investigation for wind as for subsoil, but obviously this would not be practicable or useful for a building. Subsoil strata are always there to be explored, but wind is variable. The Engineer does not generally have a choice in location, shape and size of tall buildings.

Most research workers have considered rectangular or circular shaped buildings. Can Professor Davenport or anyone else advise on Y-shaped buildings, many of which have been built in this country and abroad. One would imagine that concave face is not good aerodynamically and is likely to be subject to more wind pressure and turbulence.

The Empress State Building near Earls Court, London, is Y-shaped and is approximately 330'-0" high (see *fig.5*). There was no time available for wind tunnel tests. After the building was completed, measurements were taken during a gale of approximately 50 m.p.h. and the sway at the top was found to be less than 1/10" and the period of oscillation about 1½ seconds.

Research workers may want to know as to what information some Structural Engineers need when designing a tall multi-storey building. I would suggest a few points:

(a) Maximum wind pressure and distribution for overall stability calculations and foundation design. These should be in such a form that moments and shears can be easily calculated at any level. The effect of wind at an angle to the face of the building is also important.
(b) Characteristics of the shape and proportions of the building.
(c) Sway deformation and distortion of the building structure as a whole and also the individual members with a view to determining effects on cladding, partitions, etc.
(d) Maximum local gust intensity for cladding and its fixing to the structure.

Fig.5

(e) Dynamic and oscillatory behaviour of the building.

(f) Torsional effect due to wind at an angle and also due to horizontal variations in intensity of wind due to gusts.

(g) Stress-strain distribution in structural members. This may be important in determining the weakest point. It is not certain that one can yet rely on the results obtained from model tests. It is, however, most useful to know if there is likely to be reversal of stress at any point in vertical load bearing members due to high winds as in the case of concrete structures this could lead to cracking and high deformation with consequent detrimental effects to partitions, cladding, finishes, etc. although the building may still be structurally stable. This may not be acceptable to some Engineers. Of course prestressing could be used to avoid tension.

All the above points cannot be investigated in a single test and for important buildings three programmes may be desirable. Firstly, a solid static model could be tested in a wind tunnel for Items (a) and (b) above. This can be followed by an aero-elastic or dynamic scale model in perspex, which should reproduce relevant structural details to scale allowing for density and elastic characteristics of different materials and scale of loads, gravity, etc. The model should be clad with Molinex or similar sheeting, but care must be taken to ensure that the covering skin is not too thick and therefore likely to add to the stiffness of the model, or too thin and likely to shape and lose flat face under wind in the tunnel. This model, but without the cladding skin, can then be tested outside the wind tunnel and horizontal loads applied by mechanical means and stresses and strains measured by electric strain gauges.

A programme on the above lines for a tall building was carried out at the N.P.L. for my firm.

With different materials now being used for cladding, it would also be interesting to investigate the effect of surface roughness and relief of the structure on wind forces.

An important effect not touched on by any Contributor is "wind noise" in tall buildings. This can be very serious although it does not affect the structural stability. Double glazing appears to be desirable for higher storeys to reduce the wind noise. The shape and aerodynamic characteristics of the building are likely to affect the noise level and I wonder if by suitable design one could reduce it.

I believe ventilation of tall buildings has been studied although not referred to at this Conference. Windows have to be specially designed if they are not to be dangerous when opened in high winds. Fixed windows seem desirable in higher storeys but then air conditioning is essential. Staircases and lift shafts have to be carefully designed as they tend to act as stacks and cause draughts.

I believe Professor Mackey of Hong Kong referred to rain running upwards on the face of the building due to wind. The effect of strong winds with its vortices, wakes, eddies and turbulences on weather-proofing of building can be very important and should, I feel, be thoroughly studied.

References have been made to the effect of tall buildings on wind characteristics of the surrounding area. The writer's firm was concerned with a tall building which had to be reduced in height by two storeys to avoid interfering with the discharge of chimneys of a power station nearby.

In conclusion, I am of the opinion that research on wind effects is important and useful and must be pursued. However, the degree of reliability of test results in some cases is open to doubt and further systematic research simulating external conditions and details of models more accurately is necessary. Research although a very valuable tool cannot replace the engineering judgment and commonsense which must be an important ingredient in initial planning of research work and final interpretation of results and application in design.